Hamlyn all-colour paperbacks

Ronald Withers

# Heredity

Illustrated by John Bavosi

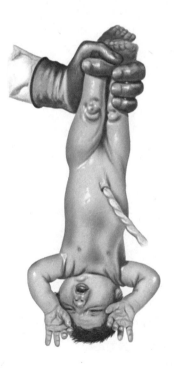

**Hamlyn · London**
Sun Books · Melbourne

## FOREWORD

There is a need to present human genetics in the context of molecular biology to educated people in the latter half of the twentieth century. The need arises from the rapid growth of molecular biology carrying with it the possibilities of changing the human genetic makeup. Already a step has been taken in the use of viruses to overcome a human biochemical deficiency. It is part of the social responsibility of a scientist to help society to understand how it could be affected by the knowledge he gains. If this book helps towards such an understanding of human genetics then it will have succeeded. If by using human genetic examples the reader is enabled to comprehend broader aspects of genetics then this is a hoped for but unexpected benefit.

R.F.J.W.

Published by The Hamlyn Publishing Group Limited
London · New York · Sydney · Toronto
Hamlyn House, Feltham, Middlesex, England
In association with Sun Books Pty Limited, Melbourne

Copyright © The Hamlyn Publishing Group Limited 1972

ISBN 0 600 00295 0
Phototypeset by Filmtype Services Limited, Scarborough, England
Colour separations by Schwitter Limited, Zurich
Printed in Holland by Smeets, Weert

# CONTENTS

- 4 Introduction
- 10 The genetic material
- 29 Chromosomes and cell division
- 44 Single character inheritance
- 57 The sex-chromosomes
- 68 Linkage
- 75 Gene action
- 86 Immunity
- 97 Mutation
- 104 Population genetics
- 124 Multifactorial inheritance
- 138 Genetics in the service of mankind
- 144 Experimental organisms
- 156 Books to read
- 157 Index

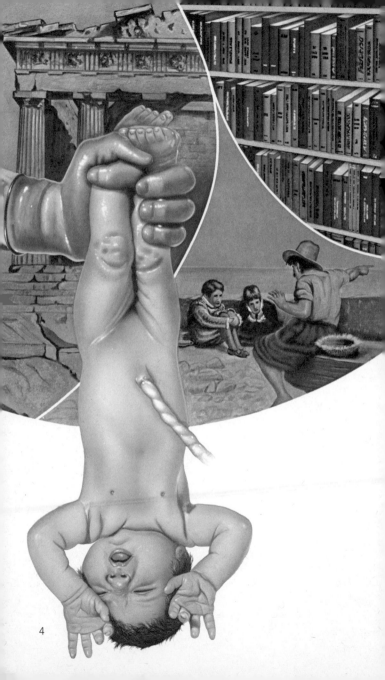

4

# INTRODUCTION

## The importance of genetics

Man has always been concerned about inheritance. Succeeding generations remain like their ancestors or differ from them for many reasons. People who are well-to-do like to be able to provide for their descendants materially; society, on the other hand, can mould its members in cultural terms. Libraries and folk-lore are among ways in which the thoughts and experiences of older generations can influence us today. We inherit our culture, our knowledge and, hopefully, some of our wisdom. The learning process is something which contributes to man's inheritance in the broadest sense of the word.

All organisms acquire, as their generations go by, some information from their ancestors. That information which has been found useful and has been selected for by the processes of natural selection is transferred from parent to child in the process of biological reproduction. It may have come to a daughter cell by the division of the parent cell, as in bacteria, amoeba and other micro-organisms, or it may have come to a little girl as the result of love, marriage, mating, and the fusion of sperm and ovum during the sexual reproduction of two human beings.

Genetics is the study of this form of inheritance. People have always been aware that genetics can affect human society. Ideas on the mixing of good and bad blood were believed by the ancients. The belief that valued qualities could be maintained by inbreeding was practised by royal houses up to a thousand years ago. Farmers have endeavoured to improve the quality of their produce by breeding. Man has seen in genetics a way in which he can maintain or change his own characteristics and those of the living world around him so as to improve his own chances of survival and those of his descendants.

## The chemical basis of inheritance

Since 1953, when Watson and Crick discovered the way in which biological information was coded, technical developments have given us a much greater command over the

The New Leicester bull *(above)* and the South Down ewe *(below)* were among the breeds developed by the late eighteenth century improvers.

*(Opposite)* Robert Bakewell (1725–1795)
Thomas W. Coke, Earl of Leicester (1754–1842)

information itself. Some of the techniques mean that we can study the information itself biochemically in a test tube; we don't have to wait for the natural reproduction of an organism to give us information by breeding. One of the most valuable tools in human genetics over the next twenty years will be the use of our cells grown in glass dishes, to provide information about what we are. Somatic cell genetics, as this technique is called, should enable us to investigate in weeks what breeding tests would take a century to discover.

Those developments mean that within the next half-century we may be able, not merely to control our environment, but to control the very material which makes us what we are. Every thinking person must be made aware of this possibility. They must understand what it is about and how it could happen. Such possibilities offer a challenge to our humanity, firstly in the literal sense and secondly in the sense of caring about what changes we all think desirable. About one in 650 children are born with severe mental deficiency –

they are called *Mongols*, or suffer from Downs Syndrome, and the genetic story of this is known. No one would be in any doubt that curing or preventing these children from suffering is a reasonable goal for human endeavour. By prevention, I do *not* mean abortion, I mean changing the information that makes them mentally defective into information which makes them normal. We could all name similar illnesses for which *genetic engineering*, as this change of information at the biochemical level has been called, would be worth while.

However, politicians do not always have humanity at heart. Perhaps we could all become blue-eyed, or all tall, or all suitably aggressive for soldiering. In genetic terms Aldous Huxley's *Brave New World* is becoming a possibility. Let me make it clear that at the moment no one has changed, in a directed way, any specific character in anything but a few bacteria. The point is that we can now see how to do it. The developments necessary are for the most part of a technical nature.

Therefore, the importance of this new knowledge must be realized and must become part of our everyday thinking. The change in emphasis will be felt in the sort of illnesses that we have. In the next twenty years, apart from illness due to accidents occurring as part of our type of society, infectious disease will become a rarity, and we will be left with a hard core of illness caused by misinformation in our genetic make-up. This will appear in childhood or will be part of the damage produced after the information is copied from cell to cell during our lifetime. Such miscopying may be part of the ageing process from which we all suffer. Understanding the nature of biological information as it passes from one biological unit to another, be it cell to cell or organism to organism, is the essence of the study of modern genetics.

*(Top)* Culture of heterokaryons (somatic cells with paired dissimilar nuclei)
*(Middle left)* Normal human cell in culture
*(Middle right)* Fibroblast from Hurler patient
*(Bottom)* Heterokaryon of cancer cell and chick red blood cell

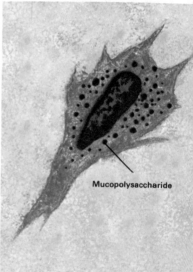

Mucopolysaccharide

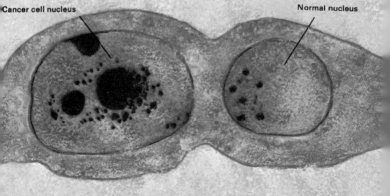

Cancer cell nucleus

Normal nucleus

## THE GENETIC MATERIAL

The search for the material which carries genetic information must start with the cell. It has been known for the past hundred years that cells consist of two major components – the *nucleus* and the *cytoplasm*. By the end of the nineteenth century the nucleus was known to contain threads of material which took a particularly active part in the process of cell division. These were called *chromosomes*. The number of chromosomes was found to be constant for a given species, whether plant or animal. In most higher organisms chromosomes could be distinguished from one another by

Internal structure of an animal cell

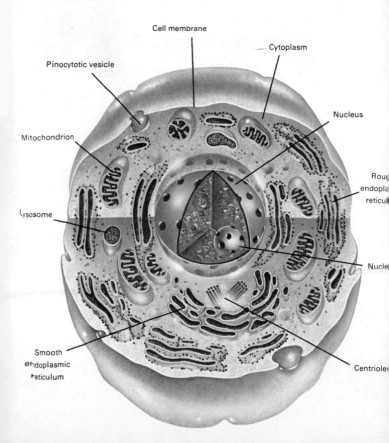

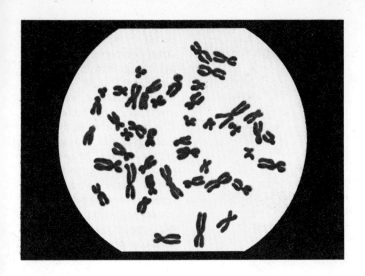

Human chromosomes – metaphase spread

their shape, and within the array of chromosomes, or *karyotype*, of the organism the chromosomes were arranged in pairs. Thus, if $N$ was the number of different shaped chromosomes, the karyotype contained $2N$ chromosomes.

It was noticed that during ordinary cell division, the chromosome number remained constant, but that in cell divisions preceding the formation of sperm and ova – the male and female *gametes* – the chromosome number was halved. We speak of the $2N$ chromosomes being the *diploid* set, and the karyotype with $N$ chromosomes being the *haploid* set. It was then realized that each parent must be contributing a haploid set of chromosomes to the offspring and that this could be a way in which information could be passed, very exactly, from parent to child. Geneticists have therefore tended to examine the nucleus with great interest as possibly containing, in the chromosomes, the carrier of genetic information.

Meanwhile cytologists studied the other non-nuclear cell components. With the advent of the electron microscope a vast amount has been discovered about cytoplasmic struc-

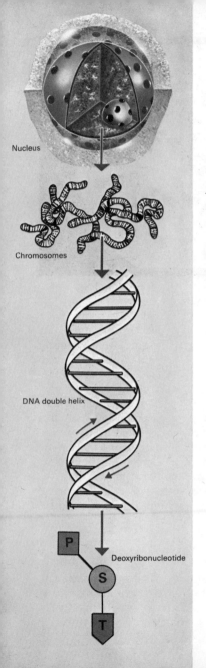

The genetic material of the nucleus

ture and function. Of especial interest was the discovery of membranes in the cytoplasm forming the *endoplasmic reticulum*. On some of the membranes were small granules called *ribosomes*. The membranes and associated granules, known together as the *rough endoplasmic reticulum*, were found very noticeably in cells which were actively concerned with the synthesis of organic chemicals – particularly proteins. Many of the proteins synthesized are enzymes which, because they can initiate or control the rate of most of the chemical reactions in the cell, are vitally important in determining what chemical reaction will occur in the cell at any given time. With the wisdom of hindsight we now realize that what a cell is like at any time depends on the chemical reactions which are occurring or have occurred in it. Therefore, the transfer of biological information between cells must relate at some stage to the production of enzymes and possibly to other cell proteins. The problem for gene-

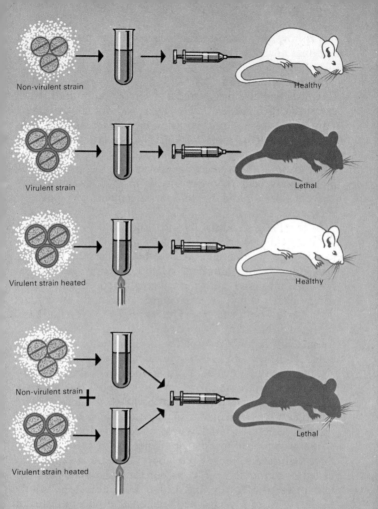

Griffiths' experiment. The death of mice injected with a combination of non-virulent and dead virulent bacteria is evidence for a transfer of information between the two strains.

tics is how chromosomes are related to the production of enzymes and other cell proteins.

## The nucleic acids

Chemists had also discovered a group of compounds in cell extracts called *nucleic acids*. They had found that there were two types of nucleic acid – *deoxyribose nucleic acid* (DNA) and *ribose nucleic acid* (RNA). Staining techniques showed that DNA was almost entirely confined to the chromosomes. Moreover, there was twice as much DNA in the body cells of organisms as in their gametes. This mirrors the diploid–haploid arrangement we spoke of earlier.

In the 1920s Griffiths injected into mice two pure bred strains of a bacterium, a *pneumococcus*. One strain had rough edges to their colonies, the other smooth. The latter was killed before injection. After extracting the bacteria from the mice and culturing the pneumococci, some of the rough bacteria had been changed to smooth by contact with *dead* smooth bacteria. The experiment was refined and repeated in 1944 by Avery, Macleod and McCarty, who found that the only part of the dead pneumococcus which could alter the living cells was its nucleic acid – in fact DNA. This was the first evidence of the ability of DNA to *transform* the information carried in a living cell to the information carried in cells from which the DNA came.

Progress was then made using viruses. These are extremely minute structures, only visible under an electron microscope, which need to exist in a living cell in order to reproduce themselves. Some viruses infect bacteria, and these are called *bacteriophages*. Phages consist only of nucleic acid and protein. Hershey and Chase were able to label the nucleic acid in a phage with radioactive phosphorus and the protein with radioactive sulphur. They then infected bacteria with labelled phage and found that only radioactive phosphorus entered the bacterium. Only the nucleic acid part of the phage is necessary for its multiplication, and the DNA must carry all the information necessary to make new phage, that is, to replicate the DNA and to synthesize the protein coat.

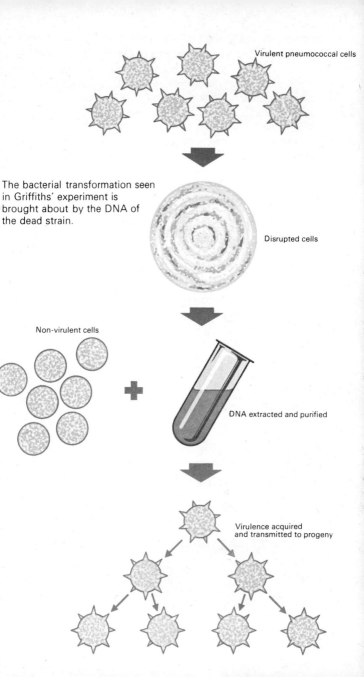

## Structure of DNA

Work therefore concentrated on DNA and its structure. Chemists found that it consisted of two types of substance – *purines* and *pyrimidines* – in equal amounts. Further analysis showed that there were two purines – *adenine* and *guanine* – and two pyrimidines – *cytosine* and *thymine*. In 1953 Watson and Crick were able to study the way in which the compounds were arranged in DNA by the technique of X-ray diffraction. Firstly they found that the four nucleotides (adenine, guanine, cytosine and thymine, each with sugar and phosphate attached) were arranged in two long chains coiled round each other to form a double helix.

Electronmicrograph of bacteriophages

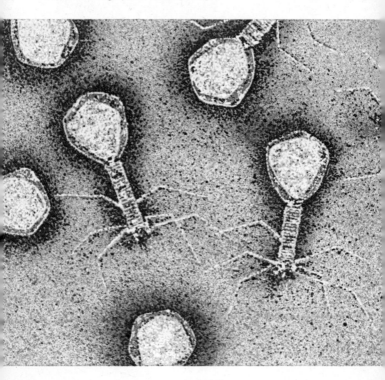

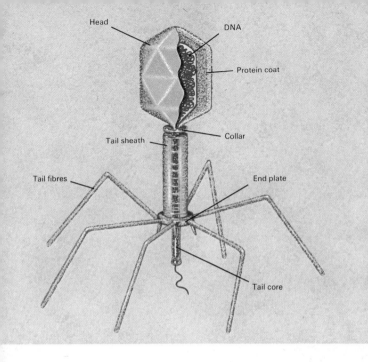

Structure of a bacteriophage. DNA is injected into the host cell where it initiates virus formation.

Moreover the chains were held together by adenine on one chain being joined to thymine on the other, and guanine on one being joined to cytosine on the other. Therefore since the four nucleotides could occur in any sequence on one chain the other chain had a complementary sequence determined by the pairing rule that has just been described.

This chemical structure was unique and it fitted what was required for a molecule 'designed' to carry information. If the helix split each chain would be able to build another chain on itself to make an exact copy of the original double helix, the sequence of nucleotides acting as a *template* on which the missing chain could be exactly built. Meselson and Stahl were able to show that this did in fact happen.

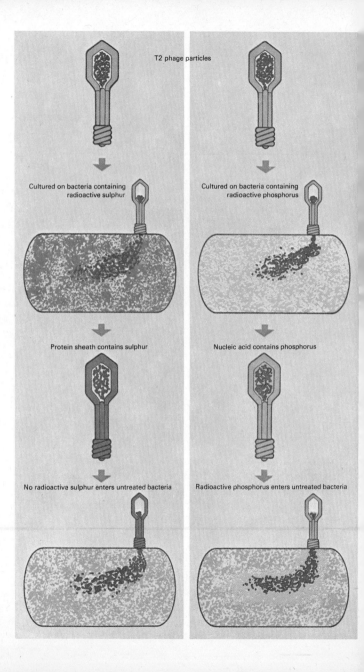

They incorporated a special form of nitrogen–$N_{15}$, a heavier isotope of nitrogen than the more common $N_{14}$– into the DNA of some bacteria so that the DNA contained only $N_{15}$ instead of normal $N_{14}$. DNA containing $N_{15}$ is very slightly heavier than $N_{14}$ and therefore a mixture of $N_{15}$ DNA and $N_{14}$ DNA can be separated by centrifuging at very high speeds into its two forms. The $N_{15}$ DNA will sink lower in a tube than the $N_{14}$ DNA when the mixture is centrifuged.

Bacteria were grown with $N_{15}$ DNA and then grown in a medium containing $N_{14}$, so that as the cells divided, if the DNA split and made new DNA the DNA should at the end of one division be half $N_{15}$ DNA and half $N_{14}$ DNA. Centrifuging would test whether this was the case. The experiment could be continued over further divisions, when the expected proportion of $N_{15}$ DNA should be as seen in the diagram. The results confirmed the expectation.

Taylor was able to show that the behaviour of the chromosomes mirrored the behaviour of DNA. He used the chromosomes of the bean *(Vicia faba)* which are easy to study under a microscope. He allowed the chromosomes to take tritium (a radioactive isotope of hydrogen) into the thymine of their DNA. Tritium will expose a photographic film, and therefore if chromosomes containing tritium are covered with a photographic film they will 'take photographs of themselves' (autoradiography). By growing the bean cells in a medium containing a tritium labelled precursor of thymine, and then transferring the beans to a normal precursor he could see what was photographed in subsequent divisions. Our illustration shows the result, which is what would be expected if each daughter chromosome *(chromatid)* contained one of the strands of the original double helix and a new complementary strand. Therefore we can now say that DNA is the carrier of genetic information and that it is mostly found in chromosomes in the nuclei of cells. To show what this means in terms of genetic information we must know how the information in

The Hershey-Chase experiment showed that only the nucleic acid components of a bacteriophage enter the bacterium.

the DNA is turned into cell chemicals, which as we saw, will be initially large molecules *(macromolecules)* of enzymes and proteins.

## Building protein molecules

Since enzymes are themselves proteins we must first look at the structure of proteins. They are macromolecules built up of units called amino acids. There are twenty essential amino acids which must be provided in the diet of the organism and come into the cytoplasm of a cell. These amino acids are arranged in a definite sequence in the protein. How does DNA inform the cell of the sequence in which to arrange the amino acids to build up proteins?

The structure of the DNA chain. The backbone of each chain consists of alternate sugar (S) and phosphate (P) molecules, the paired chains being linked together by the nucleotide bases adenine (A), cytosine (C), guanine (G), and thymine (T).

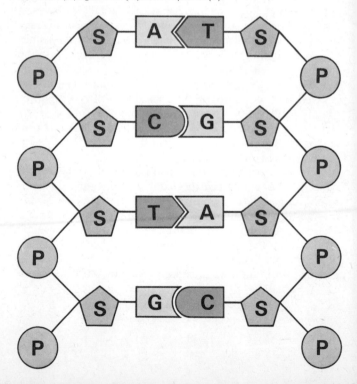

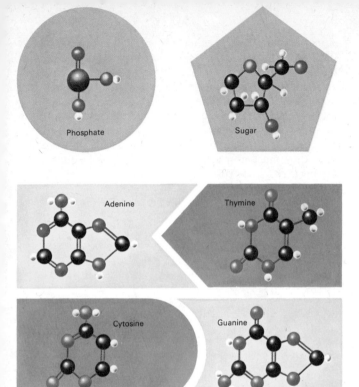

The components of the DNA chain
Nitrogen atoms—blue; carbon atoms—black; hydrogen atoms—white; oxygen atoms—red

Biochemists found that the other nucleic acid, RNA, is formed on one strand of the DNA molecule according to the same pairing rules as the complementary strand of DNA, except that a new nucleotide—uracil—is found in RNA instead of thymine. A strand of RNA is therefore built up in the nucleus on the DNA which has the complementary sequence of nucleotides as in the diagram. This strand of RNA is called *messenger* RNA (mRNA) because it takes a message, built into the sequence of nucleotides of the DNA, into the cytoplasm. The strands of mRNA go to the ribosomes on the rough endoplasmic reticulum. There

21

is in the cytoplasm another sort of RNA – *transfer* RNA (tRNA) – which has very specific properties. One part of the tRNA combines absolutely specifically with a particular amino acid while another part has a particular sequence of nucleotides which are free to pair up with any sequence present elsewhere. In fact, such sequences are available on the mRNA at the ribosomes. If the base sequences act as templates for assembling amino acids into polypeptides then they must in a sense be a code for the amino acids involved and some rule for translating the code must apply.

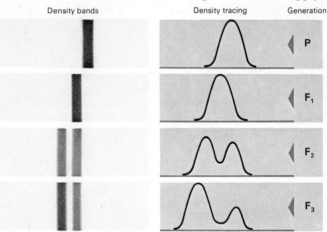

Density bands obtained at different generations by centrifugation of $N_{15}$ labelled bacteria in Meselson and Stahl's experiment

The number of items which can be coded for depends upon the length of the sequences: if the bases were transcribed in pairs, for example, then only sixteen separate amino acids could be coded for, falling short of the total number required. One of the triumphs of Watson and Crick's model for DNA was that they worked out that the sequence of nucleotides should be read in triplets, allowing 64 codable items. Subsequent work has confirmed this and a tRNA molecule with its particular amino acid has a triplet of nucleotides which can pair with a triplet in the mRNA on a ribosome. In this way building blocks of amino acids can

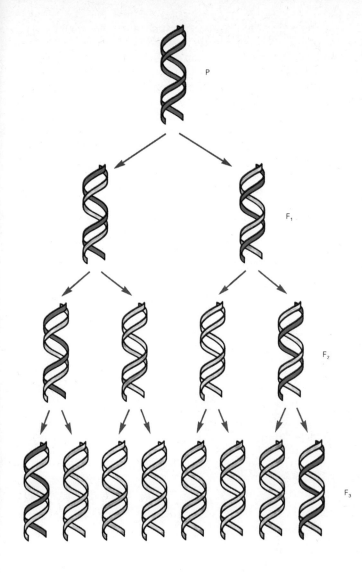

The distribution of labelled DNA (red strands) among successive generations of bacteria

be brought into sequences which are determined by the sequence of triplets in the mRNA. When the amino acids join up they form a protein (technically a polypeptide).

Thus DNA has altogether two functions. It carries a code of nucleotide triplets which can be exactly copied as the DNA goes from cell to cell during cell division. We will look at this *replication* aspect later when we describe cell division. The second aspect is that the code on the DNA can be *transcribed* onto messenger RNA which in the cytoplasm uses transfer RNA to *translate* the code into proteins. At the present the code is fully worked out. We know which triplets code for which amino acids. Some of the triplets are rather special in that they do not code for any amino acid but we know from the work done in America and England that these triplets punctuate the message. They stop the

The fate of labelled chromosomes during cell division can be followed in Taylor's experiment on the root tips of the broad bean. Sequence (a) shows the whole chromosomes at metaphase, dotted portions representing tritiated thymidine incorporated into their DNA. Sequence (b) is a diagrammatic interpretation of the results: each chromatid is shown as a pair of DNA strands, each of which subsequently replicates the other.

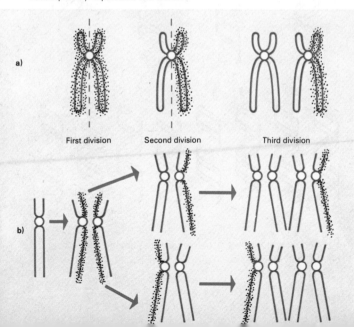

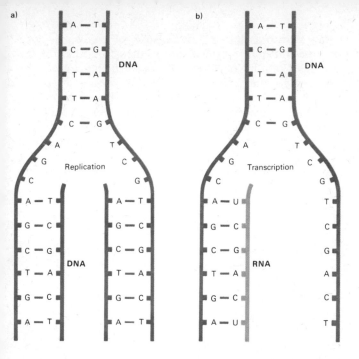

Replication (a) of the DNA sequence depends upon the complementary pairing of specific bases. In the transcription of DNA to RNA (b) the same pairing rules are followed with the exception that uracil pairs with adenine since it effectively replaces thymine.

formation of any more protein and are called *chain terminating* triplets. So that one strand of mRNA which included in the middle of it a chain terminating sequence, uracil-guanine-adenine, could make two proteins or two bits of protein provided it coded for a restart after the stop.

We have been looking at this story of the genetic code because understanding heredity in the second half of the twentieth century must have its roots in the concepts of DNA and protein formation. Whatever we say about the way in which animals and plants breed, and about the characters – blue eyes, height, milk yield, etc., which are inherited – we must refer in the end to the DNA code and the synthesis of particular proteins. The end in many instances is a long way off, but our thinking must go in that direction to understand what we are really talking about in biological inheritance.

| 1st base ↓ | 2nd base | | | | 3rd base ↓ |
|---|---|---|---|---|---|
| | **U** | **C** | **A** | **G** | |
| U | UUU Phe<br>UUC Phe<br>UUA Leu<br>UUG Leu | UCU Ser<br>UCC Ser<br>UCA Ser<br>UCG Ser | UAU Tyr<br>UAC Tyr<br>UAA STOP<br>UAG STOP | UGU Cys<br>UGC Cys<br>UGA STOP<br>UGG Tryp | U<br>C<br>A<br>G |
| C | CUU Leu<br>CUC Leu<br>CUA Leu<br>CUG Leu | CCU Pro<br>CCC Pro<br>CCA Pro<br>CCG Pro | CAU His<br>CAC His<br>CAA Glu<br>CAG Glu | CGU Arg<br>CGC Arg<br>CGA Arg<br>CGG Arg | U<br>C<br>A<br>G |
| A | AUU Ileu<br>AUC Ileu<br>AUA Ileu<br>AUG Met | ACU Thr<br>ACC Thr<br>ACA Thr<br>ACG Thr | AAU Asp<br>AAC Asp<br>AAA Lys<br>AAG Lys | AGU Ser<br>AGC Ser<br>AGA Arg<br>AGG Arg | U<br>C<br>A<br>G |
| G | GUU Val<br>GUC Val<br>GUA Val<br>GUG Val | GCU Ala<br>GCC Ala<br>GCA Ala<br>GCG Ala | GAU Asp<br>GAC Asp<br>GAA Glu<br>GAG Glu | GGU Gly<br>GGC Gly<br>GGA Gly<br>GGG Gly | U<br>C<br>A<br>G |

*(Above)* The genetic code. Each triplet of RNA bases codes for a specific amino acid or marks the end of a polypeptide chain.

*(Opposite)* Protein synthesis is effected by intracellular particles called ribosomes. Messenger RNA serves as a template onto which specific amino acid bearing transfer RNA molecules are fitted. The polypeptide chain grows as peptide bonds are formed between successive adjacent amino acids.

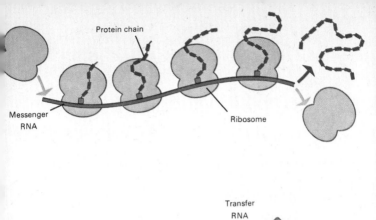
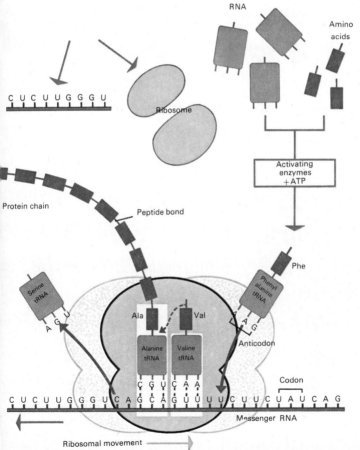

**1** Late interphase

**2** Mid prophase

**3** Late prophase

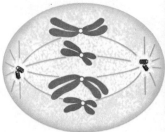

**4** Metaphase

**5** Early anaphase

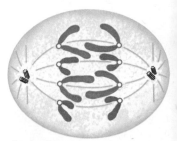

**6** Anaphase

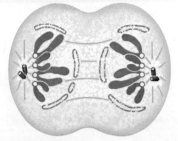

**7** Late anaphase

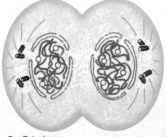

**8** Telophase

# CHROMOSOMES AND CELL DIVISION

## Mitosis

The behaviour of the chromosomes in cell division has been known for a long time and must be understood to see how DNA, carried in the chromosomes, is transferred from cell to cell. The division of body cells in animals and plants is called *mitosis*. For the majority of the life span of a cell the chromosomes are not visible using ordinary microscopic techniques. This part of the life span is called the *interphase* period of the cell. Just before cell division the chromosomes become visible as stainable threads. Each chromosome appears split and is still within the *nuclear membrane*. This is called the *prophase* stage of mitosis. Then the cytoplasm of the cell becomes organized into a system of small filaments or microtubules (the *spindle*) which radiate towards the centre of the cell from two centres at opposite ends of the cell. The centres are clearly visible in animal cells and are called *centrioles*. There is a region on each chromosome, the *centromere*, which becomes attached to the filaments of the spindle. The nuclear membrane ruptures and the chromosomes come to lie at the centre of the spindle. When this stage *(metaphase)* is reached the chromosomes are jostling around in the centre of the cell waiting for the centromere of each chromosome to split. When it does so, the split chromosomes *(chromatids)* move to opposite poles of the spindle. This stage is called *anaphase*. When the chromatids reach the poles *(telophase)* a fresh nuclear membrane is formed around them and the cytoplasm of the cell between the poles is cut into two *(cleavage)* by the cell wall and two daughter cells are formed. These daughter cells then go into interphase. In human cells in tissue culture the whole cycle lasts about 24 hours with the stages of mitosis occupying about $1\frac{1}{2}$–2 hours of this. Anaphase is relatively quick, lasting about 10–15 minutes. Cells of other organisms take varying periods for their cell cycle. Bacteria take about 20 minutes between divisions. (They do not divide by the mitotic process described above.)

Mitosis – the division of somatic cells. See text for description of stages.

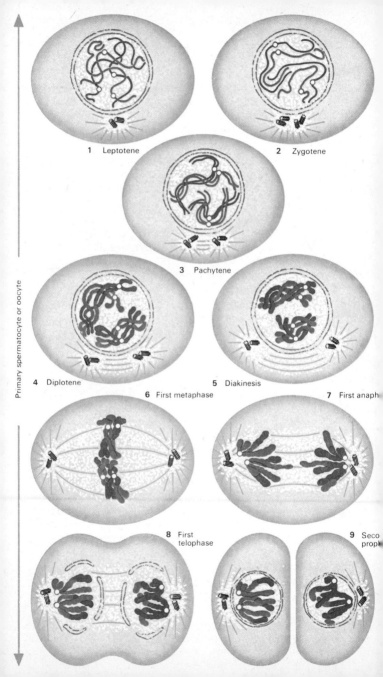

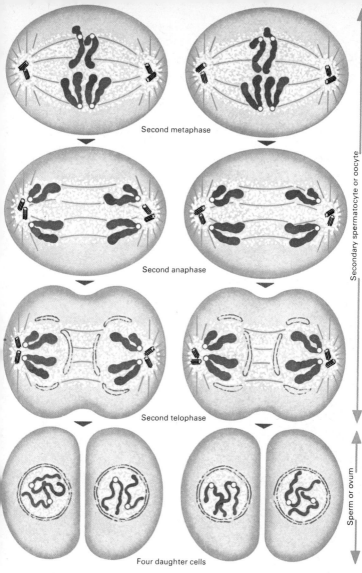

Meiosis — the formation of gametes. In the first prophase (1–5) the chromosomes pair and their chromatids cross over to form chiasmata: an exchange of segments takes place between them. The chromosomes do not split at anaphase (7), but one whole chromosome from each pair moves to each pole of the cell. The two haploid cells resulting from the first division then undergo a second mitotic division giving rise to four haploid germ cells.

Recent work has shown that DNA is replicated during the interphase, and also that RNA transcription occurs during this period. Thus the DNA content, chromosome number and presumably genetic information remain constant from parent to daughter cell as a result of mitosis.

## Meiosis – the formation of germ cells

If the chromosome number remained constant during the formation of sperm and ova we would find that when they fuse to form a new organism the offspring would have twice the amount of genetic information that the parents had. Therefore when gamete formation was examined scientists were not surprised to find that the gametes have half the genetic information of the parents. This comes about by another type of cell division called *meiosis*.

Karyotype of human male

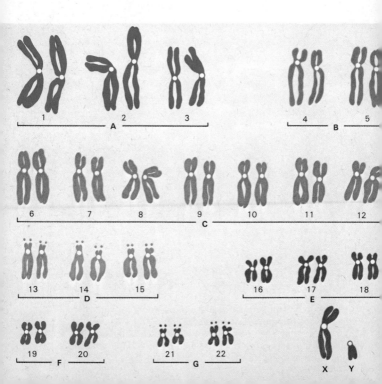

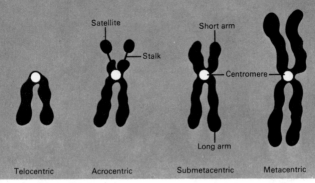

Types of chromosomes seen at mitotic metaphase. There are no telocentric human chromosomes.

This process is more complicated than we might suspect. Firstly, the chromatids appear, during prophase, to be involved in peculiar cross-like arrangements *(chiasmata* – singular *chiasma*) and secondly, as we shall see in the section on *linkage*, there is evidence that an exchange of bits of chromatids takes place between chromatids of like-shaped pairs of chromosomes *(homologues* or *bivalents)*. After this has happened, the chromosomes arrange themselves in pairs in the middle of the cell as in mitotic metaphase. In meiosis, however, the centromeres do not split. Instead whole chromosomes, with their rearranged chromatids, move to the poles. One bivalent goes to one pole while the other bivalent goes to the other pole during anaphase. There is a telophase – and it is worth counting chromosomes at this stage. If we determine the chromosome number by counting centromeres, we can see that because the latter did not split but went intact to the poles, we now have half as many centromeres in the daughter cell as we had in the parent cell. However, each centromere has two chromatids attached to it. The cell then almost immediately enters a second division during which the centromere splits and the chromatids of each move to opposite poles to form daughter cells. These daughter cells develop, in animals, into a sperm or an ovum depending on the sex of the individual carrying the cell. To summarize the effect of meiosis:

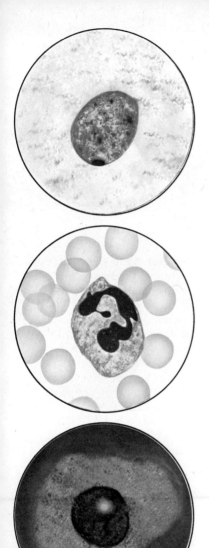

(a) It consists of two divisions, so that one cell forms in the end, four gametes.

(b) The first division effectively halves the chromosome number and the second maintains that haploid number. Gametes are therefore haploid while the parent organism is diploid. When gametes fuse a new diploid organism is formed.

(c) If we assume for the moment that two bivalents contain different genetic information, then as a result of the interchange of chromatids during prophase new chromatids are formed in which the genetic information is not necessarily the same as it was in the parents. This is the process of *recombination* about which we will say more later.

An important consequence of the formation of

Female sex-chromatin appears as a dense spot at the margin of the nucleus of a squamous epithelial cell *(upper)*, while in a neutrophil white blood cell *(middle)* it is seen as a drumstick attached to the lobed nucleus. The Y-chromosome can be identified by a fluorescence technique *(lower)*.

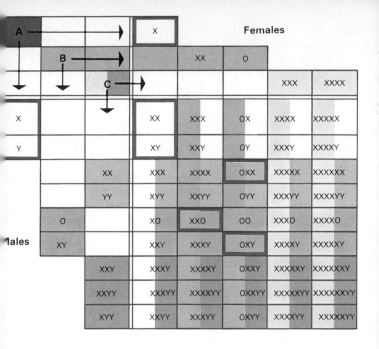

Possible sex-chromosome constitution of zygotes following non-disjunction in normal males and females. A—normal gametes.
B—gametes formed by non-disjunction at the first meiotic division.
C—gametes formed by non-disjunction at the second meiotic division.

recombinant chromosomes is that the gametes can take to the offspring the parental information arranged in different ways. Therefore offspring will not necessarily be exactly like their parents. This is one of the most important sources of variation between parents and offspring.

## Chromosome defects

In spite of pointing out that variation is derived from chromosomal rearrangements we must emphasize that these rearrangements are invisible at the microscopic level. Indeed when chromosome structure is investigated the remarkable thing is how stable it appears to be. Aberrations of the human karyotype have only been studied since

35

1956 when Tjio and Levan were able to show that human beings had 46 chromosomes and supported this finding with a technique which produced beautifully clear evidence. When other workers studied abnormal babies and children a range of abnormal karyotypes was found. Firstly when male and female karyotypes were prepared it was found that males possessed 46 chromosomes but instead of having 23 pairs of bivalents, the male had 22 pairs and two chromosomes which did not pair. This situation was similar to that found in other organisms. The odd pair are called *sex chromosomes*, one being the X-chromosome and the smaller one being the Y-chromosome. Females were shown to have two X-chromosomes. The remaining 22 pairs of chromosomes are called *autosomes*.

In 1949 Barr and Bertram working on changes in the cells of the brains of cats during electrical stimulation, explained their results by noting that a dark-staining body was frequently present under the nuclear membrane in females. This body, called the *sex-chromatin* or *Barr-body*, was

Characteristic features of Mongol palm *(left)* compared with those of normal subject *(right)*.
1 Triradius near centre, 2 Absence of pattern in thenar area, 3 Radial loop on tip of fourth digit, 4 Ulnar loops on tips of remaining digits, I Triradius at ulnar edge of hypothenar region, II Pattern in thenar area, III Whorls on tips of digits

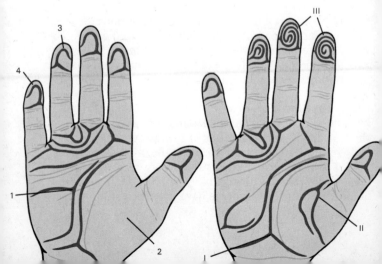

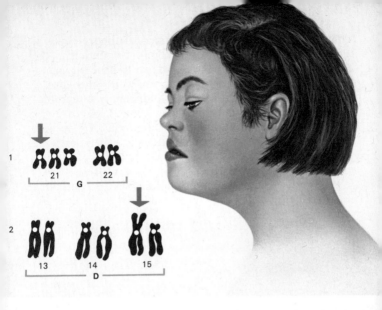

Mongolism is associated with a chromosome defect in either the D or G groups of chromosomes.

absent or very infrequent in cells of males *(sex-chromatin negative)*. This was found to be generally true of mammals, including man. However, some males were found (about one in 700) who were *sex-chromatin positive*. These men were of average to sub-average intelligence, with a penis but with very poorly developed testes, and tended to have a female body build. They showed *Klinefelter's syndrome*. On the other hand there were girls who consulted their physician because they were late in menstruating who turned out to be *sex-chromatin negative*, and showed *Turner's syndrome* – again very poorly developed gonads, short stature, and sometimes they had a fold of skin going from the shoulders up the spine to the head *(webbed neck)*. The incidence of these females was about one in 2,500. Chromosome studies showed that Klinefelter males had 47 instead of 46 chromosomes and that there were three sex-chromosomes XXY. The Turner females had only one sex-chromosome, an X, in a total of only 45 chromosomes.

Thus, for the first time in man, numerical abnormalities of the chromosome number, and therefore variation in the amount of genetic material, was shown to have an important influence on the sort of person you were. Since then, it has been discovered that Mongol children have 47 chromosómes, the extra chromosome being a number 21 (average risk one in 650). Many numerical abnormalities are now described, some with very characteristic effects, and the study of *cytogenetics*, as this subject is called, is playing an increasing part in paediatrics.

What causes numerical abnormalities? We do not know the exact cause, but we do understand part of the mechan-

Karyotype for cri-du-chat syndrome. Note short arm of one of the B group chromosomes.

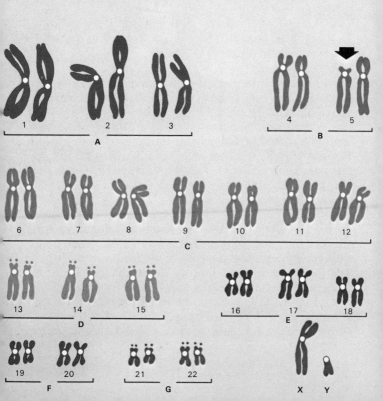

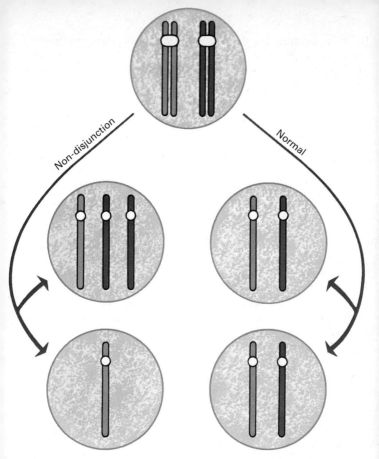

Mitotic non-disjunction in a haploid nucleus containing two chromosomes

ism. When we described mitosis and meiosis we saw that normally the chromatids of a chromosome in mitosis go to opposite poles during anaphase. If, by mistake, they were drawn to the same pole then, in a human, one daughter cell would have 45 chromatids and the other would have 47. This is called *non-disjunction*. Similarly if bivalents did not move to opposite poles during meiosis one gamete could have 24 chromatids while another gamete would have 22.

If we assume that the bivalent is an XX bivalent in a female, then some of her ova will have two X chromosomes and some no X chromosome. Fertilization of the former ovum by a Y carrying sperm will give an XXY offspring – a Klinefelter baby. The table on page 35 shows the extent to which numerical abnormalities of the sex-chromosomes could, in theory, go. From these studies we know that any individual carrying a Y is male, in terms of his outward appearance, however underdeveloped his testes are. Recently, among criminal psychopaths in prison hospitals, patients have been found with two Y-chromosomes. These men are tall and very aggressive. This raises important questions about the degree of responsibility that they carry for their crimes. As well as numerical aberrations we might expect structural aberrations of chromosomes. Some of those figured on pages 41–42 will be visible, some will not. Inversions are made visible by using autoradiography, for not all regions of the chromosome synthesize DNA at the same moment. More recently, fluorescent staining and Giemsa staining have shown banding patterns characteristic for each chromosome.

Some structural abnormalities are associated with specific medical conditions. For example, a deletion of a group B chromosome is associated with some abnormal appearance in babies and with a peculiar deformity of the larynx which causes the baby to make a sound like a cat crying when it is lying in bed. This is called the *cri-du-chat* syndrome. Another important structural abnormality is found in some Mongol children. They have 46 chromosomes, but examination of the karyotype shows that in one of the pairs of chromosomes in group D the bivalents do not match. Where the centromere should be towards the end of the B chromosome there is, instead, a short arm. This is interpreted as a translocation of a group G chromosome onto a group D, thus giving the Mongol the three group G chromosomes expected. Such children often come from young mothers who sometimes report a family history of Mongolism. The mother and father often have their

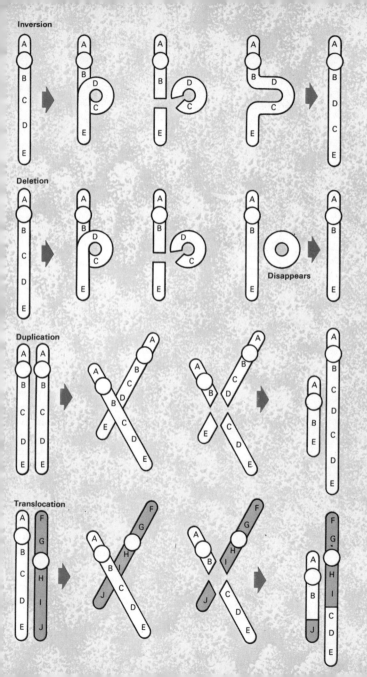

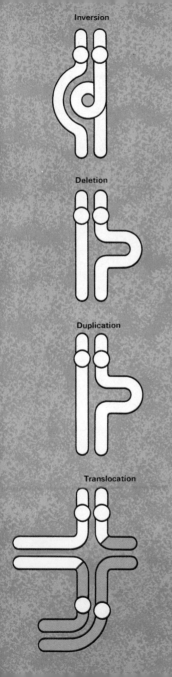

chromosomes examined, because the translocated D/G chromosome may be present in their cells – it was certainly present in one of their gametes. If it is present, say in the mother, then she only has 45 chromosomes – having only one obvious 21 chromosome (otherwise she would be a Mongol like her child). However, she carries a risk of one in three (theoretically one in four, but one combination appears to be lethal) of having another Mongol child. It is a useful exercise to try to work out, from what we have said about meiosis why the risk is what it is. If she has the chromosome in her body cells she will have received it from one of her parents and this chromosome may well be present in other members of her family. Whoever carries it has a one in three risk of bearing a Mongol child and this is why there can be a history of Mongolism in the family. This contrasts markedly with the situation when a Mongol is born with just three 21 chromosomes, for in this case there is no family his-

Chromosome aberrations apparent at meiosis during pairing

ory. Further, the likelihood of the latter occurring seems to increase with the age of the mother, rising to one in 40 for mothers of over 40.

As a result of seeing how the genetic information can be packaged into larger units – the chromosomes – and knowing something about the way in which chromosomes pass from cell to cell and from organism to organism, we will not be surprised to find, in the next section, that it is possible to make accurate predictions about the likelihood of specific conditions arising in the offspring.

The science of genetics is about the likelihood with which some aspects of biological variation are distributed in the offspring from certain matings. The next section will deal with this. Then we can see what will happen in the offspring when alterations take place at specific sites along the DNA chains.

Isochromosomes arise through mis-division of the centromere during mitosis.

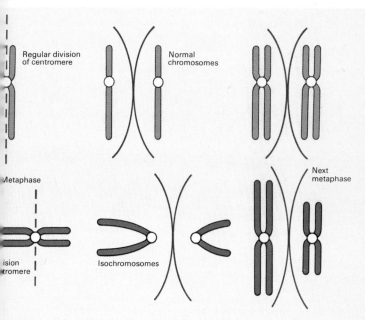

# SINGLE CHARACTER INHERITANCE

## The work of Mendel

The Abbé Mendel, in his garden at Brno in Czechoslovakia in 1866, was the first person to predict with numerical accuracy the way in which offspring differed from their parents. By counting the types of offspring obtained after mating pea plants he was able to develop the basis of the modern theory of genetics, because it led him to invent a satisfactory mechanism to explain the figures he obtained

Continuous variation in a quantitative character implies a spread of values occurring with different frequencies within a population or group of individuals. When variation is discontinuous individuals conform to one or other clearly distinct class.

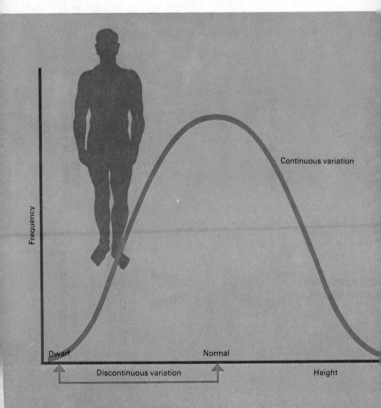

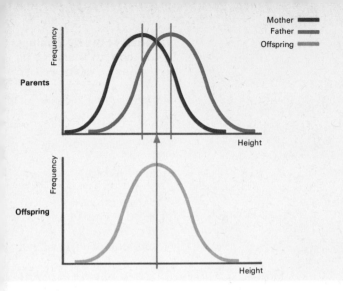

For any quantitative character, such as height, offspring tend to fall midway between their parents.

Before describing what happens we must decide what one counts in genetic experiments.

Variation in a population can refer to differences in discrete characters. Mendel's peas were round or wrinkled, tall or short. They had green or yellow seed leaves. Such variation is called *discontinuous*. As it turned out Mendel was lucky to have chosen such characters. Galton, at the end of the last century, looked at other sorts of variation such as is found in intelligence and height. This variation is *continuous* in the population, individuals with average values being commonest and the extremes being less frequent. Galton noticed that the height of offspring tended to be midway between the parental heights as shown in the figure. This led him to propose an erroneous theory of blending inheritance.

## Phenotype and genotype

The observed character that is studied is called a *phenotype*. The reader must notice that unless the phenotype shows

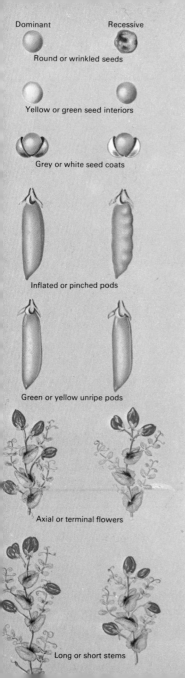

Mendel worked with characters that varied discontinuously.

variation we can say nothing about how it is inherited. Therefore, we start with at least two alternative phenotypes. When the breeding of these is examined six types of mating can be described. Let us consider the phenotype of human dwarfism (achondroplasia).

The diagram shows a need to classify dwarfs into:
(a) Those whose parents were both dwarf – calling them *dwarf 1*.
(b) Those where one parent was normal – calling them *dwarf 2*.

These are called different *genotypes* of the same *phenotype*. Furthermore, in this example we notice that in mating 3 the *normal* phenotype has disappeared in the offspring. *Dwarf* is said to be *dominant* to *normal*, and *normal* can be said to be *recessive* to *dwarf*. In practice we make reference to the *normal*, so that the alternative character is said to be dominant or recessive as the case may be. In this case *dwarf* is dominant. However, although the *normal* phenotype disappeared in *dwarf 2* individuals, when they mate, individuals who are *normal* comprise a quarter of the offspring. As we shall see, this led Mendel to realize that there was no blending of characters and that the mechanism of heredity was particulate.

We have already seen that each individual is diploid, that is for each type of chromosome there is a pair of homologues. What Mendel called 'factors' we now call *allelomorphs* (more commonly, *alleles*) and each individual carries two alleles, one at a specific place (a *locus*) on each of a pair of homologous chromosomes. The word *gene* was invented by Johannsen (1911) to denote a unit character.

Types of matings which can occur for a pair of alleles governing achondroplastic dwarfism.

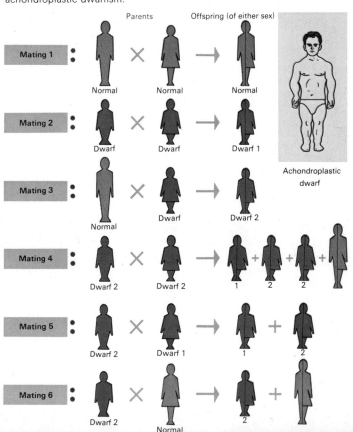

Achondroplastic dwarf

This could be present in at least two alternative forms *(alleles)*. However, there is a tendency to talk about genes as being the same as alleles. The important thing to remember is that a specific locus on a chromosome is always related to a particular variant of a character, and what determines that variant is the allele (or gene) occupying that specific locus. In the example given, mating 1 describes the result of the mating of individuals carrying two *normal* genes *(d)* at the loci concerned with height on a pair of bivalents. Mating 2 describes the mating when the individuals carry two genes *(D)* which give the *dwarf* phenotype. Because a gamete contains the haploid set of genes, the *normal* individuals *(dd)* produce only gametes carrying one *d* gene; the *DD* dwarfs produce only *D* carrying gametes. Therefore, the *dwarfs* produced by mating 3 must be carrying one *D* gene and one *d* gene, and can be described as *Dd* individuals. When this pair of loci carry the same sort of gene the individual is said to be *homozygous*. We have, therefore, two types of *homozygous* individuals—*DD (dwarf 1)* and *dd (normal)*. Characterizing the individuals in this way is to describe their genotype. There is yet another genotype— *Dd (dwarf 2)*, which is said to be *heterozygous*. Now we can write out the six matings in genotypic terms:

|  |  | Offspring: |  |
|---|---|---|---|
| Mating: | | Genotype: | Phenotype: |
| 1. | $dd \times dd$ | $dd$ | *normal* |
| 2. | $DD \times DD$ | $DD$ | *dwarf* |
| 3. | $dd \times DD$ | $Dd$ | *dwarf* |
| 4. | $Dd \times Dd$ | $\frac{1}{4} DD + \frac{1}{2} Dd + \frac{1}{4} dd$ | $\frac{3}{4}$ *dwarf* $+ \frac{1}{4}$ *normal* |
| 5. | $Dd \times DD$ | $\frac{1}{2} DD + \frac{1}{2} Dd$ | *dwarf* |
| 6. | $Dd \times dd$ | $\frac{1}{2} DD + \frac{1}{2} dd$ | $\frac{1}{2}$ *dwarf* $+ \frac{1}{2}$ *normal* |

We must notice that the most informative mating is No. 6. It is called a *back cross*. It is informative because it reflects the proportions in which the gametes are produced in the heterozygote. As we saw in the previous chapters we would expect an individual *Dd* to produce at meiosis *D* gametes and *d* gametes in equal proportions. Since the *dd* individual

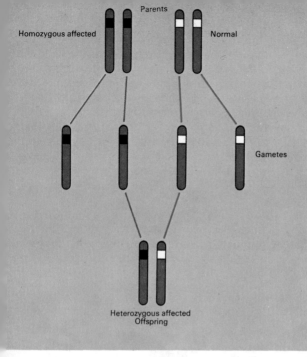

In a mating between a homozygous dominant and a homozygous recessive all the offspring are affected heterozygotes.

produces all *d* gametes and these do not show in the zygote because of the dominance of the *D* genes which pair with them, the ½ *dwarf* : ½ *normal* ratio of offspring shows that the *Dd* individual did in fact produce ½*D* gametes and ½*d* gametes. This means that the *D* and *d* genes did not blend, but remained as discrete genes which appeared again segregated during gamete formation.

In the example given, an abnormal trait was dominant to the normal. About 1,000 such conditions are known in man. They include such conditions as cataract, some forms of deafness, the ability to taste certain bitter substances like phenyl thiocarbamide (PTC) and the ability to roll one's tongue at the sides. However, in some 500 or 600 other cases it is the abnormal character which is recessive and the normal which is dominant. The commonest abnormal gene in Great Britain is the gene causing a disease in children called *cystic fibrosis* of the pancreas, and this is recessive. In such a case, both parents are, to all appearances, normal and yet one in four of their children can be affected with cystic fibrosis. Such children come from mating 4, in which both parents are *heterozygotes*. We can see from this that one of the aims of medical geneticists is to be able to detect those apparently normal people who might be heterozygotes or *carriers* of the disease gene.

## Dominance

There are some common misconceptions about what is meant by dominance and recessiveness. Some people imagine that dominant genes are common, or that in some way they dominate over recessive genes. Originally the

Transmission of genes is determined by chromosome behaviour. In a mating between heterozygotes the effect of the dominant gene (black) appears in three-quarters of the offspring.

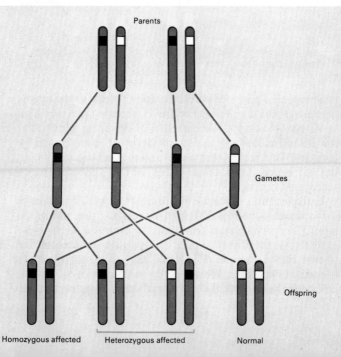

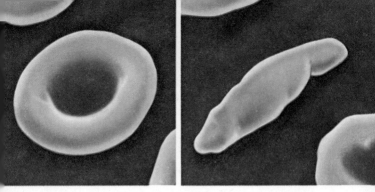

Normal red blood cells *(left)* and cells from a sickle-cell anaemia patient *(right)* seen under a scanning electron microscope.

word was used to describe a condition in which the phenotype of one homozygous genotype was expressed as the phenotype of the heterozygous genotype. This meant that the recessive gene, which was present in the heterozygote appeared to be doing nothing towards producing the phenotypic character of the heterozygote. This was found not to be always true. Many cases were found where the heterozygote had a different phenotype from either homozygote. Such genes were said to be *intermediate* or *co-dominant*. An example of this was the gene pair at the locus which determined whether one had a disease, common in some tropical countries, known as *sickle-cell anaemia*. Individuals heterozygous at this locus were *not* normal, especially under conditions of oxygen shortage, but neither did they have sickle-cell anaemia. Under a microscope, their red blood corpuscles were sickle-shaped as in the picture. This did not make them very ill (or die, as people do with sickle-cell anaemia), but they certainly were not normal. They had *sickle-cell trait*.

Our understanding of dominance and recessiveness was advanced when this disease was investigated. It was found that people suffering from sickle-cell anaemia made only one sort of haemoglobin – the red pigment found in red blood cells. This sort of haemoglobin was called haemoglobin S. Normal people had haemoglobin A. When the haemoglobin of people who had sickle-cell trait was investigated it was seen that they made both haemoglobin A and haemoglobin S. Now haemoglobin is a protein, and from what we said about DNA and protein synthesis, we can see more of what the sickle-cell gene does. It produces haemoglobin S instead of haemoglobin A. The phenotype has therefore changed from being a description of the symptoms of sickle-cell anaemia to being a description of a type of protein whose presence causes the symptoms of the disease. Protein biochemists are able to separate the constituents of the proteins they study. There is a technique known as chromatography whereby, once a complex compound has been broken down into its constituents, its components can be separated by allowing them to run on a piece of blotting paper in a suitable solvent. The constituents run for different distances from the point at which they are dropped onto the blotting paper. When this was done for the haemoglobin S it was found that one amino acid – glutamic acid – was not present in haemoglobin S, but was present in haemoglobin A. Further, valine (another amino acid) was present in haemoglobin S and not in haemoglobin A. Other studies showed that this difference occurred at the sixth amino acid along the $\beta$-chain of haemoglobin. The substitution of valine for glutamic acid was the only difference between the two haemoglobins. Moreover, haemoglobin is a complex protein consisting of four subparts or *polypeptides*. There are two pairs of identical polypeptides called $\alpha$- and $\beta$-chains. Haemoglobin A and haemoglobin S have identical $\alpha$-chains. Thus the phenotype for the gene has become even more precise. The genes differ in that the haemoglobin gene normally codes for glutamic acid in the sixth triplet of the sequence *(cistron)* coding for the $\beta$-chain, but codes for valine in the corresponding sequence when the haemoglobin S gene is present. Workers on the genetic code know that a cytosine-thymine-cytosine triplet codes for glutamic acid, and a cytosine-adenine-cytosine triplet codes for valine. Therefore, difference in DNA at one base pair – substituting an adenine for a

Blood smears from sickle-cell patient *(above)* and normal subject *(below)*.

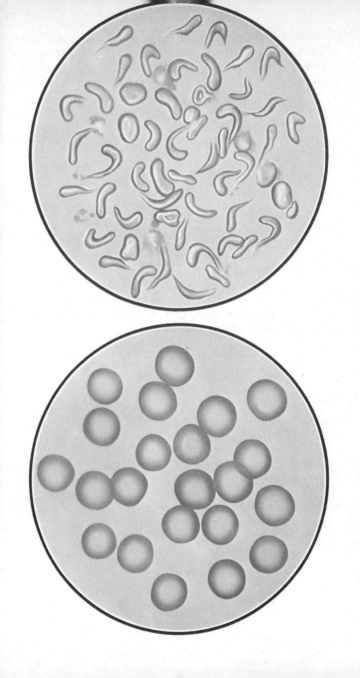

thymine – is sufficient to cause the gene difference we know as the 'gene for sickle-cell anaemia'.

The sickle-cell allele codes for a particular sequence of amino acids in the $\beta$-chain. There are other variants of the $\beta$-chain, for example haemoglobin E which has the amino acid lysine instead of glutamine in the twenty-sixth position. This variant, and the others, are alleles to haemoglobins S and A. Therefore, although the codes for the alleles differ in different triplets, the proteins produced (that is, the $\beta$-chains) are alternatives. Thus alleles (or genes) regulate the production of a particular protein, or part of a protein, and this is the genetic unit that we study. Biochemists may find, as we saw for haemoglobins A and S, that the differences between the alleles may be specified down to base pair changes, but functionally it is the polypeptide (the part of the protein such as the $\beta$-chain of a haemoglobin molecule) that we are concerned with when we refer to a gene allele. From what we have said it should be obvious that when we are in a position to describe any phenotype in protein terms the concept of dominance and recessiveness should disappear. However, some diseases are known which appear

Diagrammatic representation of human haemoglobin chains showing positions of some variants in the amino acid sequence

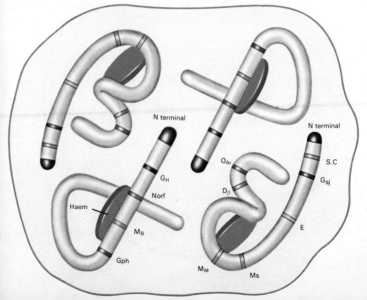

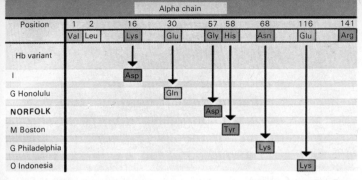

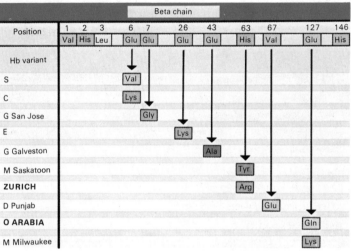

Haemoglobin variants and the amino acid substitutions corresponding to them

to be due to the lack of an enzyme which speeds up a specific chemical reaction. Enzymes possess *active sites* which play a part in the reaction. Now diseases caused by the lack of enzymes are usually genetic in origin and often recessive. This means that the individuals who are heterozygous have one gene which codes for and therefore produces the enzyme and another gene which does not. Because the enzyme is produced this makes the other gene 'recessive'. However, we have to remember that the enzyme is charac-

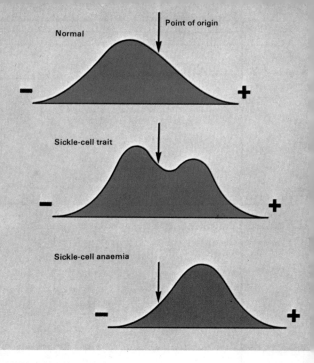

A technique like chromatography (electrophoresis) separates *whole* haemoglobins in an electric field.

terized by its ability to speed up a reaction. Therefore, what we have said is that the recessive gene product does not *behave* like the usual enzyme. We have not said that there is no product of the recessive gene. Therefore, to say that it is recessive is to confess our ignorance of what it produces.

Since genes produce proteins, then with suitable techniques we should be able to find the protein products of both genes in a heterozygote. When we do this, however, both genes will be co-dominant. This is why human geneticists are trying to find the gene products of recessive genes. If they succeed then it will be possible to detect heterozygous individuals *(carriers)*. Genetic counselling can then be given to these individuals, for if two persons are carriers then one in four of their children will be affected.

## THE SEX-CHROMOSOMES

In the chapter on the chromosomes we saw how human males carry a Y-chromosome as well as an X-chromosome, and how females carry two X-chromosomes. We have just seen that genes occupy loci on chromosomes and that there are six ways in which matings produce variation among offspring, assuming that the loci involved are on autosomes. We might expect that they can also occur on the X- or Y-chromosomes. The distribution of genes occupying such loci will be affected by the behaviour of the X- and Y-chromosomes in meiosis.

As can be seen from the diagrams there are three possible places where loci are available on the X- and Y-chromosomes:

1. On the part of the X-chromosome for which there is no homologous part of the Y-chromosome. Such a gene is said to be *X-linked*. This is by far the commonest situation and such genes are said to be *sex-linked*.

Regions of the sex-chromosomes showing possible types of sex-linkage

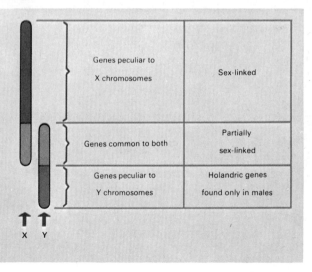

2. On the part of the Y-chromosome for which there is no homologous part of the X-chromosome. Such genes are said to be *Y-linked*, and the trait is said to be *Y-borne*.
3. There may be parts of both the X- and Y-chromosomes which are homologous. Such genes would be said to be *partially sex-linked*. There is considerable doubt as to whether there are such loci in man. It is very difficult to detect such genes.

Sex-linkage is evident in a cross between a Rhode Island Red cock and a Light Sussex hen. In birds the male is XX and the female XY so the dominant white gene on the hen's X-chromosome is passed on to the male in the next generation.

Light Sussex hen and Rhode Island Red cock

## Y-linked genes

At the moment only one such gene is known. It produces long hairs on the lower part of the ear lobe, and has been described in several populations in India by Dronamraju. What will be the characteristics of such a trait? In a normal mating we expect the sex ratio of males to females to be equal. If a gene is on the Y-chromosome then the sons will be affected, and only males can be affected.

## X-linked genes

Firstly the genes occupying loci on the X-chromosome can be dominant or recessive. Most are recessive and this has certain consequences. Colour blindness is a condition caused by such a gene. Secondly a male $X^cY$ will be affected because, although he can never be homozygous for the recessive gene he will never be able to be heterozygous for it, because he will not have two X-chromosomes. If a colour-blind man marries a normal woman the mating is:

|  | Female: | | Male: |
|---|---|---|---|
| Parents: | XX | × | X$^c$Y |
| Gametes: | X | | $\frac{1}{2}$X$^c$ + $\frac{1}{2}$Y |
| Offspring: | $\frac{1}{2}$XX$^c$ | + | $\frac{1}{2}$XY |

In other words, all the sons will be normal, but all the daughters will carry the colour-blind gene on the X-chromosome they inherited from their father. Now what happens when these carrier women marry normal men?

|  | Female: | Male: |
|---|---|---|
| Parents: | XX$^c$ × | XY |
| Gametes: | $\frac{1}{2}$X + $\frac{1}{2}$X$^c$ | $\frac{1}{2}$X + $\frac{1}{2}$Y |
| Offspring: | $\frac{1}{4}$XX + $\frac{1}{4}$XX$^c$ | $\frac{1}{4}$XY + $\frac{1}{4}$X$^c$Y |

Since the father transmits an X-chromosome to his daughters and a Y-chromosome to his sons, a Y-borne trait such as hypertrichosis must pass exclusively from father to son and appear in each generation.

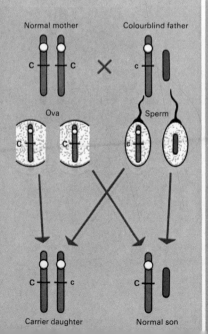
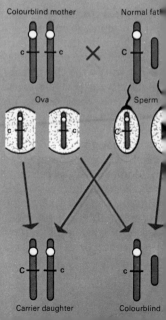

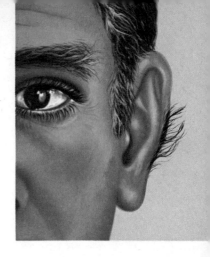

Hypertrichosis *(right)* is a Y-borne trait. The pedigree *(below)* of a family in which the trait occurs bears out the male to male transmission.

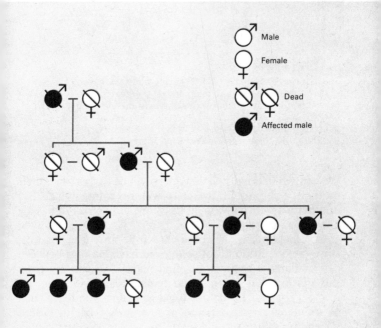

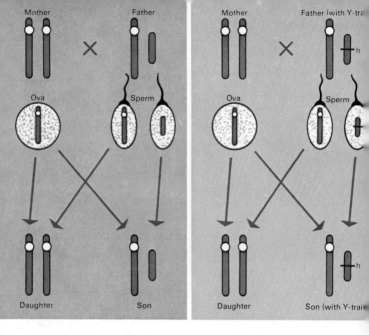

The mode of transmission of an X-borne trait (sex-linked) can be seen in the inheritance of colourblindness. Where the father is colourblind and the mother normal, none of the offspring are affected but the daughters are all carriers. Where the mother is affected (homozygous recessive) and the father normal the daughters are carriers and the sons are affected.

We can see that half the daughters will be carriers like their mother, and half of the sons will be colour-blind. Put another way, such a woman has a one in two risk of having affected sons.

Colour-blindness has been chosen because it is possible, since the condition is not serious, to have matings between carrier females and colour-blind males. We can see from this mating that colour-blind females can arise who are homozygous for the colour-blind gene. In a condition such as *haemophilia* (the absence of factors which clot blood) which is more serious – certainly more serious under medical care as it existed thirty or more years ago – the affected males frequently died before they reproduced and homozygous females died in puberty when they started to menstruate.

## Inheritance of sex-linked characters

We can now see several characteristics of sex-linked inheritance:
1. It is usually, and far more frequently, males who are affected.
2. If affected males do mate with normal females, none of their sons are affected but all their daughters are carriers. Since the gene is recessive, this means that its phenotype disappears in that generation.
3. The carrier females transmit the gene to half their sons. The phenotype may therefore appear again. This is

When a colourblind father mates with a carrier mother half the sons are normal and half affected while half the daughters are carriers and half affected.

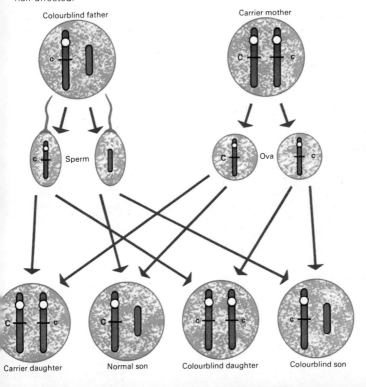

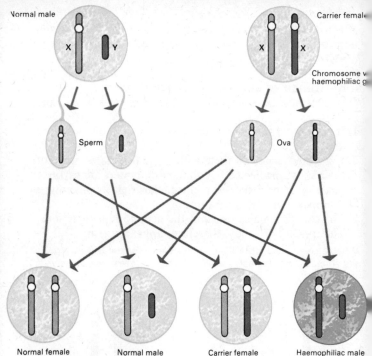

Haemophilia, another sex-linked defect, is inherited in the same way as colourblindness. Haemophiliac females do not survive to bear children, however.

sometimes called 'criss-cross' inheritance. The character misses a generation.

Many people seem to think that missing a generation is characteristic of all genetic behaviour. We hope you will now understand that this is only true if a gene is located on the X-chromosome. Such genes include other conditions such as the commonest type of *muscular dystrophy*, a severe wasting condition of the muscles, and *favism* in which serious anaemic effects follow the eating of the broad bean, *Vicia faba*, and which is relatively common in some parts of the Mediterranean and Far East.

Genes can be dominant or recessive or co-dominant, and can be on autosomes or on sex-chromosomes. Although one finds that a gene producing a given phenotype may be commonly dominant, the same phenotype may be produced rarely by another gene which is recessive or even sex-linked. Therefore, each family has to be carefully examined to see the way the gene is behaving in *that* family.

There are two further points to be made when considering sex-linkage. There are genes, like the one which produces frontal baldness, which are carried on autosomes and which are therefore equally likely to be present in either sex, but which only produce the phenotype when male sex hormones (androgens) are present. This means that only males will be affected. Such genes are said to be *sex-limited*, and can be confused with *sex-linked* genes. Careful examina-

Sex-limitation is seen in the expression of the gene for baldness.
b – normal gene; B – gene for baldness.

| Genotype | Androgen | Oestrogen |
|---|---|---|
| b  b | not bald | not bald |
| B  b | bald | not bald |
| B  B | bald | not bald |

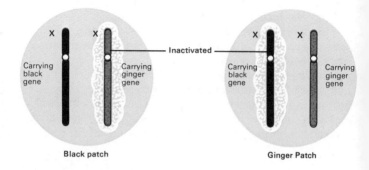

tion of a pedigree is necessary to distinguish the two situations. Another instructive exercise is to work out what happens to dominant and recessive genes which are *sex-limited* to either sex. You will then discover how the way in which they segregate differs from the way in which sex-linked genes segregate.

This is a convenient place to point out that if we consider the amount of protein produced in a homozygous cell by two X-borne genes, that is in an XX cell, and compare the amount produced in a male cell XY by its one X-borne gene, we might expect female cells to produce twice as much as males. In the majority of cases where this has been tested it has been found that this does not happen. If you think about it, it might mean that women were 'twice as much something' as males. The findings mean that a mechanism for compensating for the dose of genes present *(dosage compensation)* is necessary.

Dr Mary Lyon, several years ago, put forward the theory that when more than one X-chromosome was present, all but one of the X-chromosomes are randomly genetically inactivated during development in every cell – so that in any given cell and its descendants in the organism only the genes on one particular X-chromosome can function. This theory has been tested, both cytologically and genetically, and has found considerable support. Exactly what is meant by 'inactivation' is not known. She explains the patchiness of the colours (which are determined by the sex-linked genes) of female tortoiseshell cats by 'inactivation' of the X-chromosome in the different cells which subsequently divide to form a patch of cells, all producing hair of the same colour, because the gene on the other X-chromosome has been 'inactivated' in those cells. It follows that if you know of a tortoiseshell male cat, it must at least be XXY. Very few have been found.

---

According to Lyon's hypothesis, during the development of a cell one of its X-chromosomes is inactivated. This same chromosome will be inactive in any of that cell's descendants. The patches of a female tortoiseshell cat may arise because of a random inactivation in cells giving rise to patches of hair.

# LINKAGE

We are now in a position to consider what will happen when the genes at two loci segregate. The test situation is when an individual is heterozygous at both loci, for if at one locus the genes are $A$ and $a$, and at the other locus $B$ and $b$ the individual's genotype could be written as $AaBb$. When such an individual was mated to an $aabb$ (the back cross mating) then one would expect, on what we have said so far, the following types of offspring:

|  | $AaBb$ | $\times\ aabb$ |
|---|---|---|
| Gametes: | $\frac{1}{4}AB + \frac{1}{4}Ab + \frac{1}{4}aB + \frac{1}{4}ab$ | $ab$ |
| Offspring: | $\frac{1}{4}ABab + \frac{1}{4}Abab + \frac{1}{4}aBab + \frac{1}{4}abab$ | |
| Phenotypes: | $\frac{1}{4}AB + \frac{1}{4}Ab + \frac{1}{4}aB + \frac{1}{4}ab$ | |

One must remember when thinking about the types of gametes that no gamete can have both $A$ and $a$ or $B$ and $b$. The genes at each locus go into different gametes, but may go with either of the genes at the other locus. This is what is meant by the *independent assortment* of genes. The different possible kinds of gamete should appear in equal propor-proportions if the genes are randomly paired.

Over fifty years ago, Morgan was studying the way in which the offspring were obtained in a cross between white-eyed, yellow-bodied male fruit flies *(Drosophila)* and red-eyed, grey-bodied female *Drosophila*. At the eye colour locus, *red eye (R)* is dominant to *white eye (r)*, and at the body colour locus, *grey body (G)* is dominant to *yellow body (g)*. The offspring turned out to be of four sorts:

white-eyed with yellow bodies;
red-eyed with grey bodies;
white-eyed with grey bodies;
red-eyed with yellow bodies.

Since the males were recessive at both loci, they only produced gametes with the genes for white eyes and yellow bodies. Therefore, the females were giving four different types of gamete. In one experiment he had 2,205 offspring

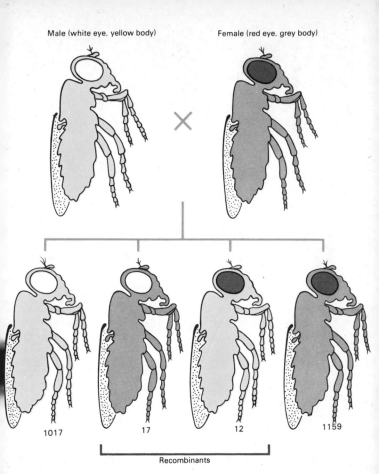

The result of Morgan's classic experiment on recombination in *Drosophila*

and from what we said about independent assortment the four types of offspring should be in equal proportions, that is to say there should be about 450 of each type. In fact he obtained the results shown in the illustration. Morgan was able to explain this as being due to the two loci being on the same chromosome – the X-chromosome in this case – of the fruit fly.

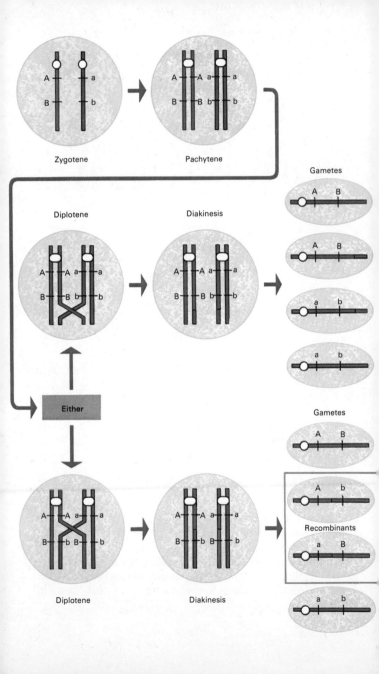

## Crossing over

If you refer back to the pictures of meiosis on pages 30–31 you will notice that exchange of chromatid segments occurs during prophase. The illustration on page 70 shows what happens if the exchange or 'crossing over', as Morgan called it, occurs outside the two loci and also between the the two loci. The percentage $29/2176 \times 100 = 1.3\%$ could loci are together the less likely it is for a cross over to occur between them. The net result is an excess of gametes in which the genes occupying the loci are arranged in the same way as they were in the parental pre-meiotic cell. The gametes containing genes arranged in the different way produced by crossing over are called *recombinant* gametes. Looking now at the experiment we can tell:

1. That the genes affected by these particular eye colours and body colours are on the same chromosome (the X-

*(Opposite)* The mechanism of recombination. Exchange of chromatid segments takes place during meiosis.

*(Below)* The frequency of recombination between genes at different loci on the same chromosome pair is an indication of the distance between the loci. It is thus possible to map the positions of loci in relation to each other.

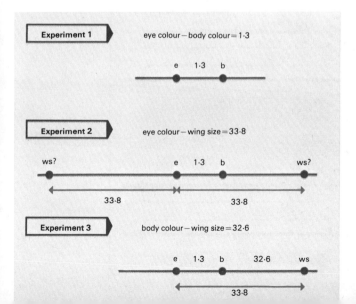

chromosome) as Morgan spotted.
2. That in the female doubly heterozygous flies, of the two X-chromosomes, one had the white-eyed gene and the yellow body gene occupying loci on it, while the other had the red-eyed and the grey body gene on it.
3. As a result of crossing over between the two loci, two rearrangements produced as recombinant gametes, white-eyed and grey-bodied gametes, and red-eyed and yellow-bodied gametes.

## Gene maps

In 1923 Sturtevant pointed out that since the proportion of offspring formed from recombinant gametes was affected by the distance between the loci involved, this recombinant proportion could be used as a measure of distance between the two loci. The percentage $29/2176 \times 100 = 1.3\%$ could be used as a distance measurement. By finding similar results between eye colour and wing size where the value

The cis-trans complementation test distinguishes mutations occurring in the same or in different functional genes.

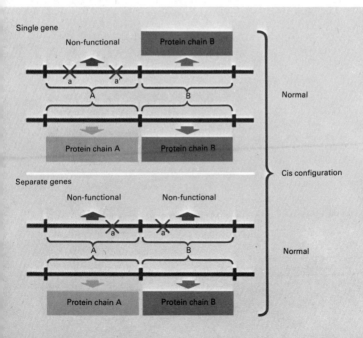

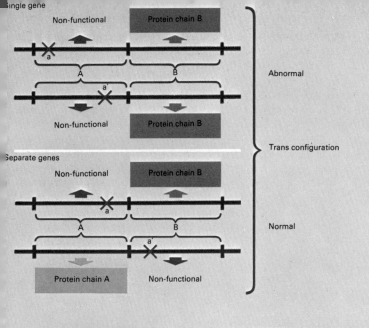

was 33·8%, Sturtevant showed that one could position the three genes on a chromosome in the order shown in the diagram. Since then similar breeding experiments have enabled us to map genes in linear order in many organisms. Groups of genes which give this sort of result are said to be *linked*, and such groups are called linkage groups. Indeed, if we are right in assuming that genes are on chromosomes there should be the same number of linkage groups as the number of *haploid* chromosomes. Workers have found this is so in all organisms that have been studied.

Why is it worthwhile investigating linkage, apart from the purely genetic interest? It could be very helpful in forecasting the likelihood of people getting a disease. If a particular blood group gene is *linked* in a family with a gene which causes disease, then we could predict from the blood group of a child whether it is likely to have the disease. In those countries where therapeutic abortion is allowed one could then prevent a seriously malformed or diseased child being born, or at the very least could provide, in advance, all necessary medical facilities.

## The cistron

Over the last twenty years scientists have realized that bacteria, with a generation time of twenty minutes, can be bred by the hundreds of millions in a single test tube or on a petri dish filled with agar jelly, and any recombinant can be left to grow until it forms a colony big enough to see. In this way the power to resolve genetic loci has improved so much that microbial geneticists can resolve genetically down to the base pair level of coding that biochemists find as a result of protein analysis. This is called *fine gene analysis*. Such techniques enabled Benzer, in 1957, to define a gene in terms of the functional unit which made a polypeptide (part of a protein). He first used the term *cistron* to define this unit and his work was based on what he called the *cis-trans* test. If genes are arranged so that the two dominants are on one chromosome and the two recessives on the other they are said to be in *coupling* or in *cis*. Whereas when a chromosome has the dominant gene at one locus and the recessive at the other and the reciprocal arrangement on the other chromosome, the genes are said to be in *repulsion* or in *trans*. Benzer found that when the phenotype was a polypeptide – that is when one considers a gene as a functional unit – variants which did not give the polypeptide when they were in *trans* were parts of the same gene. It will be a long time before genetic analysis can get to this level of resolution in man. However, geneticists have found that human cells in tissue culture might be able to show their genetic nature. In the last few years a technique using hybrid cells has led to the location of human genes on particular chromosomes. Lines of Chinese hamster cells have been developed which are deficient in certain enzymes such as thymidine kinase. It is possible to fuse these cells with normal human fibroblasts. The hybrid cells will grow because the genes in the human cell supply the enzyme. The human chromosomes are lost as the hybrid divides. When chromosome 17 is lost the cell dies. The human gene producing thymidine kinase is thus located.

# GENE ACTION

Up to now in this book we have assumed that whenever an organism carries a particular gene the effect of that gene will show itself. We have assumed that this is so whether the effect is the presence of a particular protein or whether we are taking as a phenotype a character like the red eye in *Drosophila*. The illustration shows a human condition called *lobster claw* which is inherited as a dominant. Individuals carrying the gene for lobster claw would be expected to have both hands and feet severely deformed. In fact, however, they may only have one hand affected, or one foot. (This is called variability of *expression*.) Why then does the gene not show itself in all the sites that it influences?

Pedigree of *lobster claw* involving an affected female and her offspring by three different fathers. There is a considerable variation in skeletal structure among the affected children.

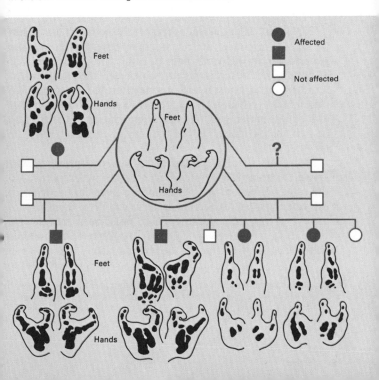

There are some genes which do not express themselves at all in individuals who are known to carry them. From looking at a pedigree in which such individuals occur, the number of times that the gene does express itself can be worked out and compared with the number of times that it should express itself. The percentage of times it does express itself is called the *penetrance* of the gene. When doctors study human genetics they sometimes talk about a gene being dominant, with variable expression and low penetrance. On thinking about this one realizes that this is covering quite a lot of ignorance about how the gene really works, so one has to be careful before using these terms. Then again the variation in expression is sometimes a variation in *kind* of expression instead of variation in *amount* of expression. There is a human gene which affects the colour of the outer wall of the eye turning it blue rather than the normal white. Some people with this gene *blue sclerotic* also suffer from very fragile bones. Yet others are deaf. Some 40 per cent have all three conditions. This is what is meant by a difference in *kind* of expression.

The situation underlying this sort of picture is that genes act on the organism during its development. Indeed some genes act only relatively late in adult life. We often talk about a gene causing a disease which has late age of onset. Huntington's Chorea is such a condition. The average age of onset is thirty-five years. It is caused by a dominant gene and people can marry, have children, and hand the gene on to half their children before they know that they have the disease.

However, the differentiation of an organism from a single cell to an adult must be a story of getting cells into the right position at the right time for genes to be switched on, causing a cell to become a nerve cell or to form part of an eye of the right colour. There must be a method for regulating gene action. The red eye gene of *Drosophila* cannot work when there is no eye. Similarly if a cell is in

The gene determining the condition *blue sclerotic* may affect the sclera of the eye (1), the limb bones (2), and the bones of the ear leading to deafness (3).

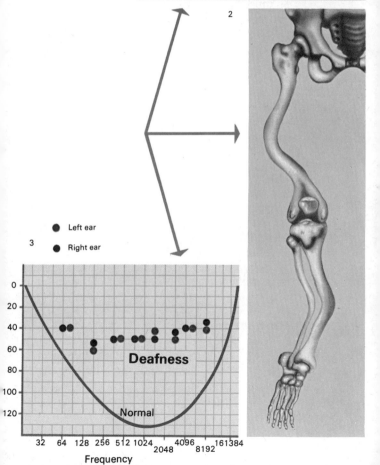

the wrong place or its genes are switched on at the wrong time, then development subsequent to this might also be abnormal just because of the early defect. Therefore abnormalities can be due either to basic malfunction of a gene or to the effects this has on subsequent development. This latter effect is called the *pleiotropic* action of a gene. Whenever a geneticist sees a difference in *kind* of expression he suspects that none of the kinds he is looking at is the basic gene action. The *grey lethal* gene of the mouse shows how pleiotropy can work. The geneticist then starts to look for the primary gene effect – and from what we have described this must be traceable to some protein defect and research moves in this direction.

We may consider also the way in which a chromosome defect produces a condition such as Mongolism or Klinefelter's syndrome. Through non-disjunction an extra chromosome is incorporated into the karyotype and produces an abnormal phenotype. Yet certain chromosome abnormalities can occur in living organisms with little ill effect. While most higher plants and animals are diploid ($2N$), in some plants, for example cultivated irises and many lilies, the *entire* set of chromosomes is duplicated resulting in a tetraploid state ($4N$). These plants reproduce themselves in the $4N$ state and appear to function normally. It would seem that the difficulty lies in the balance of chromosomes. In diploid, haploid or polyploid cells with a full complement of chromosomes, gene products may be formed in varying amounts but in a nevertheless balanced condition. A single additional chromosome disrupts the balance. Genes do not act independently of one another in determining the phenotype: the interaction of their products is critical.

## Control of gene action

Research is also going on to investigate what it is that switches a gene on at the right time. In higher organisms we do not know, but in 1961 Jacob and Monod discovered the

The immediate effect of a gene may lead to a complex sequence of interrelated events.

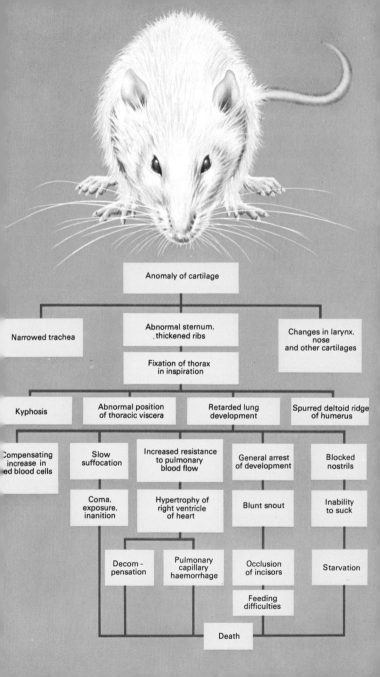

way in which this is done in bacteria. They subsequently received a Nobel Prize for this discovery.

They were studying the production of a group of enzymes in the bacterium *Escherichia coli*. They found that even if the genes which coded for the enzymes were present in the cells, the bacteria did not produce the enzymes unless certain substances were present in the culture medium. When these substances (often substrates for the enzymes) were present they induced the formation of the enzymes. The production of specific enzymes is itself a phenotypic character and such characters are called *inducible enzyme systems*. One such enzyme was $\beta$-galactosidase and the inducer necessary in the medium was $\beta$-galactoside. Mutant bacteria appeared for which the inducer was no longer necessary. Such mutants were called *constitutive* mutants because the bacteria always produce the enzyme. This mutation could be shown to occur at a different site from the genes (called *structural* genes) which produced the enzymes and since it sometimes controlled the working of the structural genes it was called a *regulator* gene. Jacob and Monod postulated that the regulator gene could produce a protein (a *repressor* substance), which normally repressed (switched off) the action of the structural gene, but that it could also combine with the *inducer* and when it did so it no longer acted on the structural genes, which being derepressed (switched on) produced their messenger RNA, which in turn coded for the enzyme. In the system on which Jacob and Monod worked more than one enzyme was induced by the inducer. Since it was difficult to see how more than one structural gene could be switched on by one repressor and one inducer, they postulated another gene (the *operator* gene) on which the repressor worked. This operator gene then controlled messenger RNA formation from each structural gene in the system. Since they postulate the site of repressor action to be a gene, mutants of the operator gene should be available. These were found – one mutant, for example, was unable to recognize the repressor and therefore kept the structural genes in a state of active RNA

Postulated scheme of enzyme synthesis in a bacterial operon

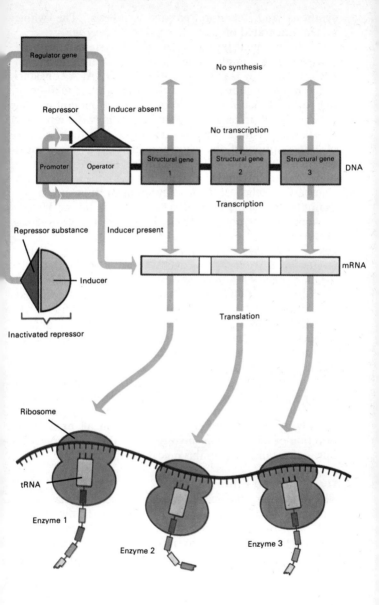

synthesis and therefore enzyme synthesis. The whole system illustrated on page 81 is called an *operon*.

At this moment several other operon systems have been discovered. A complete operon – the *lac* operon – has even been successfully isolated as a strand of DNA. A considerable amount is known of the way in which the regulation of this sort of structural gene works in terms of mutant regulator and operator genes, the chemistry of how the repressor substance recognizes the operator locus, and the sequence of transcription of the structural genes. Therefore most geneticists are tempted to extend the operon concept to higher organisms, and to imagine that this is the way in which genes are switched on during development. It is known, for example, that some hormones can act as inducers. Unfortunately we have not found in higher organisms the mutants of the regulator or operator genes which would provide the genetic evidence that they exist in higher organisms. Moreover, for a switching mechanism to work successfully the messenger RNA must be very short-lived. As Harris and his colleagues at Oxford have shown, there is evidence that this cannot be the case in some higher organisms. It has been known since 1926 that the alga *Acetabularia* is able to regenerate a cap which has morphological features depending on the genes present in its cell nucleus even at several weeks after the nucleus has been removed. This suggests that the nucleus transferred messenger RNA molecules to the cytoplasm before it was removed, and these molecules were long-lived. Thus there are still unresolved problems of how genes are actually switched on in all organisms. Solution of the questions involved will add considerably to our understanding of embryonic development.

## Cytoplasmic inheritance

Before leaving gene action, we must remember that information can be passed from cell to cell via the cytoplasm. The most obvious way in which cytoplasmic inheritance might be suspected in a given system is that transmission appears to follow the maternal line. This is because the sperm of most organisms contain very little cytoplasm, whereas the ovum has large amounts of cytoplasm.

Examples of cytoplasmic inheritance are relatively few and quite diverse. One of the earliest to be described is the control of the direction of coiling of the shells of a snail *Limnea*. The illustration shows that the genotype of the female parent determines the direction of coiling in the offspring. The angles of the first division of the egg are determined by the female parent and are preset by the maternal genotype in the cytoplasm of the egg.

Another example of cytoplasmic inheritance is found in plant variegation. The offspring of a cross between a normal plant and a variegated plant are either normal or variegated depending solely upon the character of the maternal line. The cytoplasmic particles concerned with variegation are the *chloroplasts* which contain the photosynthetic pigment *chlorophyll*. The inheritance of variegation appears to depend upon the transmission of the chloroplasts themselves. If a plant were to lose its chloroplasts, the nucleus

Transplantation experiments with the single-celled alga *Acetabularia* reveal the overriding role of the nucleus in influencing growth. However, the ability of the isolated tip to form a cap indicates the presence in the tip cytoplasm of messenger substances (presumably messenger RNA).

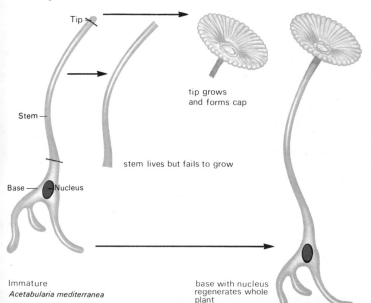

would be unable to synthesize further chloroplasts, although nuclear genes are able to influence their characteristics. Like genes, chloroplasts undergo mutation and mutant chloroplasts in turn give rise to mutants. It is the transmission of these colour mutant chloroplasts which accounts for the inheritance of variegation.

Some of the cellular DNA is to be found in the mitochondria, so that these cytoplasmic organelles which contain enzymes necessary for the respiratory activity of the cell have their own DNA control for these enzymes. Of special interest is the fact that some cytoplasmic particles which are passed from cell to cell during division are able to become combined with nuclear DNA in certain circumstances. These *episomes* are found, for example, in bacteria. One such factor, the F factor, when incorporated into the bacterial chromosome confers on the bacterium a high frequency of recombination which is very useful in studies on the genetic analysis of bacteria. Since episomes frequently take with them bits of bacterial chromosome as they migrate from chromosome to cytoplasm this helps in the analysis. Moreover, cytoplasmic inheritance implies that the episomes behave as if they were infective or transmissible agents. A special group of episomes called R factors are responsible for transferring to *Salmonella typhimurium* (an organism responsible for gastro-enteritis, especially prevalent among babies in hospitals) resistance to the action of antibiotics which normally cure the illness by killing the *Salmonella*. There is now considerable evidence that particles which confer *infectious drug resistance* are very dangerous to human health. Not merely do they confer resistance in pathogenic organisms to one antibiotic, but they collect resistance to several other antibiotics as they pass from bacterium to bacterium during conjugation. This constitutes a considerable disease hazard as drugs are rendered useless by the accumulation of resistance in the organisms which cause diseases.

The determination of shell coiling in *Limnea* is an example of cytoplasmic inheritance. The maternal genotype influences the direction of coiling via the cytoplasm of the egg cell.

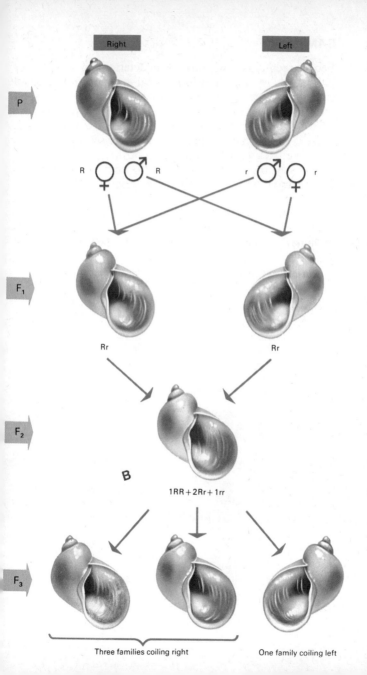

# IMMUNITY

Because genes play a major role in determining what sort of individuals we are, both in physical and mental terms, we can say that genes determine what self is. On the other hand the self, in order to preserve itself, must be able to recognize what is not self. We can easily do this for events outside ourself, by touch and sight and our other senses, but when we come back to ask the question of things inside us, it is not so easy to give an answer. Yet the body is able to recognize material which is foreign to it very successfully.

## Antigens and antibodies

When foreign protein enters the blood, the body responds by making a chemical which attaches to the foreign protein

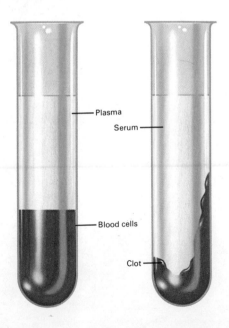

The constituents of blood. If blood is prevented from clotting *(left)*, blood cells separate out from the clear plasma which contains the clotting elements. When blood clots *(right)* the clotting elements are removed from the fluid leaving serum.

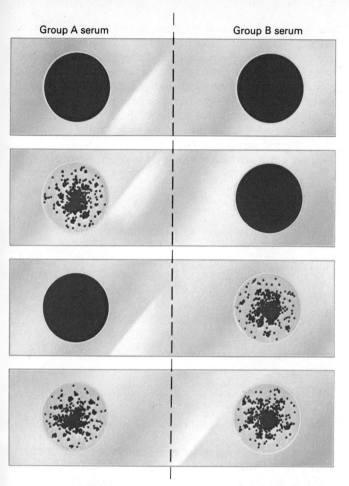

The effects of serum A and serum B on blood cells is used to determine blood groups. Tests are carried out on glass slides. Serum A, containing anti-B, agglutinates both B and AB cells. Transfusion of B or AB cells into a group A recipient would therefore stimulate an anti-B reaction. O cells, with no antigens, would not be reacted to.

and forces it to go out of solution, or by providing cells which ingest it, and which themselves can be removed from the body. This is what happens during inflammation or when a part of the body goes septic. We call the foreign protein an *antigen*, and the chemical produced in response to it by the body, an *antibody*. The antigen-antibody reaction is called an *immune response*. This is not the place to detail all the findings of immunology, which is the study of the immune mechanisms. But just as the genes code for the self, they must code for making antibodies and for the ability to recognize self and not self.

Most, if not all, cells in the body produce antigens which

| The ABO blood group system | | |
|---|---|---|
| Individuals belonging to | have on their cells | have in their serum |
| Group A | Antigen A | Anti-B |
| Group B | Antigen B | Anti-A |
| Group AB | Antigen A + Antigen B | Neither antibody |
| Group O | Neither antigen | Anti-A + Anti-B |

Antigens and antibodies of the ABO blood group system

are part of what is meant by self. The most obvious ones are found on red blood cells and are characterized as the blood groups. All through history various people had tried to transfuse blood from person to person or from animal to person, sometimes with disastrous results because the transfused blood was recognized as non-self and was rejected, blocking the blood vessels of the host and killing the person being transfused. Such blood was *incompatible* with the recipient's blood. In 1900 Landsteiner discovered the reasons for the incompatibility. He found that on the surface of red blood cells there occurred two types of antigens, which he called A and B. In serum (the straw coloured liquid which remains after all the cells in blood have clotted out) two antibodies were found which he called anti-A and anti-B which agglutinated red cells carrying A and B antigens respectively. These antibodies were probably produced early in life when, for example, an individual with A-carrying cells was challenged by something (often bacteria) which carried a B-like antigen. Because the body was A it recognized B as not self and produced anti-B in response to it, and vice versa.

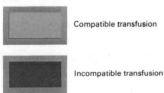

Compatibility relationships within the ABO system

It turns out that there are *three* genes at this blood group locus. One gene produces A antigen, one produces B, and one produces neither – it is called O. (In fact, the gene produces O substance before A or B antigen is synthesized and if A or B genes are present O substance is made into A antigen or B antigen.) A is therefore dominant to O, and B is dominant to O, but if the individual is heterozygous for A and B, both substances will be produced. There are, therefore, four phenotypes, A, B, AB, and O, and six genotypes AA, AO, BB and BO, AB, and OO. Matings can therefore be worked out:

|  | Male: | Female: | Phenotypes: |
|---|---|---|---|
|  | AO × | BO | A × B |
| Gametes: | $\frac{1}{2}A + \frac{1}{2}O$ | $\frac{1}{2}B + \frac{1}{2}O$ |  |
| Offspring: | $\frac{1}{4}AB + \frac{1}{4}AO + \frac{1}{4}BO + \frac{1}{4}OO$ | | $\frac{1}{4}AB + \frac{1}{4}A + \frac{1}{4}B + \frac{1}{4}O$ |

Blood groups, of which more than twelve different systems are known, are used to decide cases of paternity, as well as being important in matching for transfusion.

Another important blood group system is that concerned with the *Rhesus factor*. For simplicity, we can regard people as being either Rhesus positive ($Rh^+$) or Rhesus negative ($Rh^-$). In fact, the situation involves three very closely linked loci, but the details need not bother us. When a mother is Rhesus negative, we can regard her as being homozygous recessive. If her husband is Rhesus positive then, depending on whether he is heterozygous or homozygous, either half or all of the children will be *Rhesus positive*. The mother will treat the embryo's red blood cells as foreign and will make an anti-$Rh^+$ antibody which will agglutinate the blood of the embryo. This only happens when, or if, there is a leak of the embryo's blood into the mother's circulation. In the first pregnancy, if everything has gone well, a leak is only likely to happen at birth, so

---

The risk of haemolytic disease in pregnancies where Rhesus incompatibility is involved increases through successive pregnancies *(upper)*. If anti-$Rh^+$ formation is stimulated by leakage of the embryo's blood into the maternal circulation the mother's antibodies will pass across the placenta into the circulation of the embryo *(lower)*.

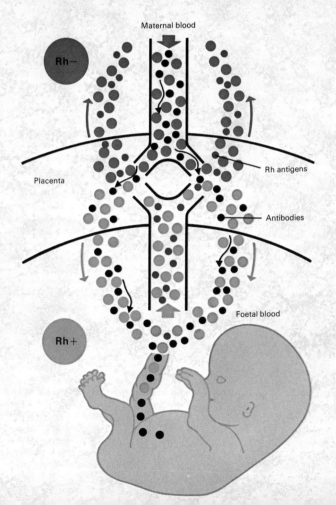

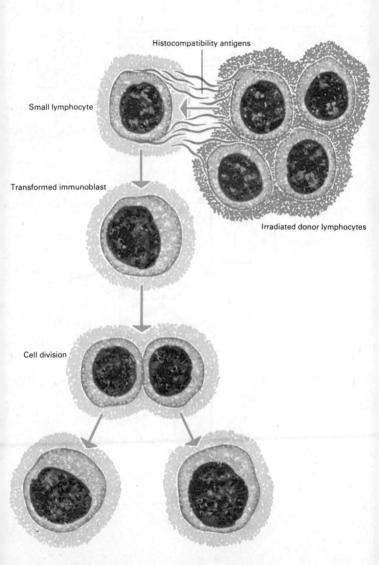

that the first child may not be affected. However, as a result of the birth, the mother may start to build up anti-Rh$^+$ antibodies, which may be of sufficient strength to attack the blood of the second child. It is even more likely to happen for subsequent children.

This condition has been a common cause of an illness, haemolytic disease of the newborn, in the past. The later children of Rh$^-$ mothers were often born very jaundiced, but in recent years an exchange blood transfusion has often saved the life of the child. Very recently Professor C. A. Clarke of Liverpool University conceived a brilliant way of solving the problem.

It does not matter if a Rh$^-$ male is immunized against Rh$^+$ blood. Such males can therefore be used to obtain a supply of anti-Rh$^+$ antibody. If this is injected into the Rh$^-$ mother it can be used to eliminate any Rh$^+$ cells from her baby that get into her circulation *before* she has time to build up her own anti-Rh$^+$ antibody. Since ABO incompatibility confers some protection against Rhesus incompatibility, the tests that have been done have been with mothers where there is ABO compatibility between the mother and child.

## Tissue rejection

More recent developments in immuno-genetics involve the recognition problems which arise when tissues from other people or organisms are grafted onto or into a host organism. With the advent of advanced surgical techniques for liver, kidney and heart transplants geneticists have been faced with trying to understand the genetic basis of the rejection of the grafted organ which can occur. The fact

---

Histocompatibility can be assessed by mixed lymphocyte culture reaction. Donor lymphocytes are killed by irradiation rendering them unable to react but leaving their histocompatibility antigens intact. These antigens are recognized by normal recipient lymphocytes which react in varying degrees by enlarging and dividing, thus indicating the extent of incompatibility between donor and recipient. The fate of a transplanted organ correlates with the severity of the reaction.

that the body recognizes whole tissues as foreign shows that the tissues have antigenic determinants. Part of the difficulty has been in identifying these tissue or transplantation antigens. Luckily, some progress has been made by finding that the white blood cells also carry antigens which may represent those found in other tissues. These findings involved analysing by computer the reactions of many sera to a group of white cells. This work was initiated in Holland by Van Rood in 1962; subsequently workers from all over the world have met periodically to test groups of sera against groups of cells. As a result three systems of linked genes have been found. These are called *histocompatibility* genes and human studies so far confirm the sort of system found in the mouse. From the latter studies a number of rules can be stated, but it must be remembered that they only apply to inbred strains–and human beings are not inbred. An inbred strain of animals has all the genes in common within the strain.
1. Grafts within strains succeed.
2. Grafts between strains fail.
3. Grafts from either parent, when they are of different

Human skin has been successfully transplanted into a mouse.

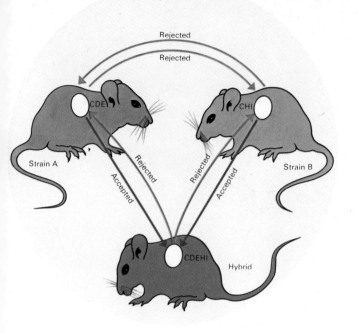

The fate of skin grafts exchanged between parents of different strains (and different transplant antigens) and their hybrid offspring.

strains, succeed in the offspring, but grafts from offspring to parent fail.
4. If the offspring are mated together, then grafts from their children succeed in the original offspring.
5. However, grafts from either inbred strain to grandchildren are accepted by some grandchildren and rejected by others.

It will be a long time before human studies can be so precise. The evidence, however, is strong that there is a genetic basis to this. At the present, genetic studies are not too helpful and therefore surgeons are relying on ways of preventing the immune reaction taking place by using immunosuppressive drugs.

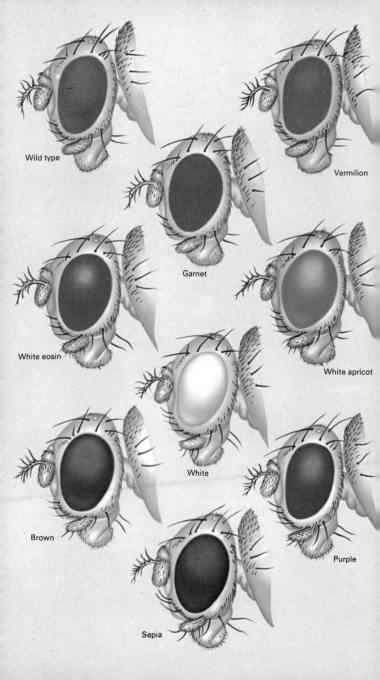

## MUTATION

All through our study of genetics we have seen that variation and its inheritance is what genetics is about. How is the variation produced in the first place? It is produced by *mutation* of an original gene. Mutation appears as a sudden change in a phenotype that is unexpected in the sense that the changed phenotype has not been produced as a result of segregation during meiosis. In other words, if one has a pure bred line of an organism, occasionally an organism is produced which differs from the rest of the line.

If one is breeding pink tulips occasionally a plant produces a flower with yellow stripes in it. The plant might be a mutant. We will know whether the variation is a mutant or not by whether, when we breed from the mutant, its characteristic behaves as we would expect it to do on genetic theory. This excludes sudden environmental change or infection as the cause of the variation. Mutations can be produced in two main ways:
1. Structural and numerical chromosome changes, such as deletions, inversions and polyploidy.
2. Gene (point) mutations where there is a change in the DNA code.

What can cause either of these two processes? From what we know of DNA we can see that factors which affect accurate replication of the code could produce changes in sequence or changes in the actual base which occurs in a triplet. If, for example, we grew an organism in a medium which was deficient in thymine (a base) we might expect errors to occur where thymine was required in the code because of the shortage of thymine in the medium. In practice, the requisite sequence of code would not be made, and the cell would probably die. But there are chemicals such as proflavine which do not kill the organism, yet produce mutations by causing additions and deletions of bases and therefore shifts in the code. Other chemicals cause changes in specific bases and alter the coding in this way. In other words, a coding change is probably the basic mechanism

Mutant eye colours in *Drosophila*

for a point mutation. Mutations which involve structural chromosomal changes also involve DNA. The changes may even be due to properties of chromosomal constituents other than DNA.

## Mutation rate

When scientists seek to study mechanisms they find it useful to be able to measure something about their material. Hence, real progress was made in the study of mutation when Muller discovered that although mutation was rare, it nevertheless occurred at a fixed rate for a given character and a given organism. When this rate is known we can investigate the causes of mutation by studying what factors

*(Left)* The six haploid chromosomes of a pollen grain.
*(Right)* The same chromosomes after exposure to mustard gas.

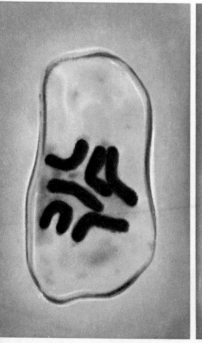
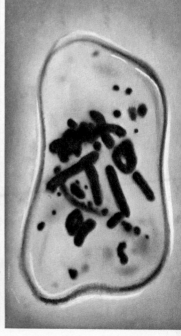

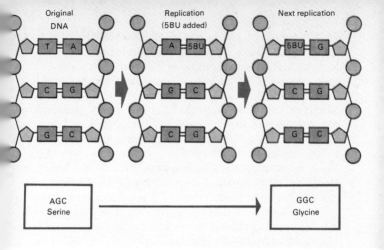

5-bromo-uracil, a powerful mutagen, may become incorporated into the base sequence of a replicated DNA chain. At the next replication, because 5BU pairs with guanine, this results in a permanent change in the code.

change the rate and under what conditions. This has given geneticists an insight into the mechanism of mutation. Because natural spontaneous mutation was rare, the discovery that X-rays will induce mutation (that is, they are *mutagenic* and increase the mutation rate) has led to a study of induced mutation. By inducing mutation we have more experimental control over the process. Other mutagens are now known and these include various chemicals, such as mustard gas, some antibiotics and certain dyes like acridine orange, and other forms of radiation such as ultra-violet light.

How can we compute the mutation rate? We can examine an entire population, such as was done in Denmark, and find, say, the number of dwarfs (those with chondrodystrophy). As we saw in our exposition of Mendelism this is a dominant condition. Therefore, the rare individuals (actually 10 in 94,075) will be heterozygous for this dominant gene. Now two of the dwarfs came from marriages of dwarfs, but the remaining eight came from normal parents. They therefore represent mutants, i.e. eight mutants in

94,075 people or nearly one in 12,000. Now each of these mutant individuals represents one mutant gene or a pair of loci, so altogether there are 24,000 genes involved at that locus in the 12,000 individuals, of which one has mutated. The mutation rate is, therefore, one in 24,000 genes. Mutation rates have been calculated in other organisms for other genes and they all have incidences ranging from one in $10^5$ to one in $10^8$. We must notice that mutation can occur in any cell of an organism. However, in practice it is most likely to occur in the dividing cells, because the change in the DNA code usually occurs during DNA synthesis; it is the newly synthesized DNA which is different and carries a different code from the strand from which the copy is made. If a mutation occurs in the ordinary cells *(somatic cells)* of the body, as opposed to the germ cells (ova or testes) then a somatic mutation results. A somatic cell with changed DNA may go on dividing and may start a tumour. However, it will not be passed on to the offspring of the individual

The relationship between mutation rate and radiation dose.

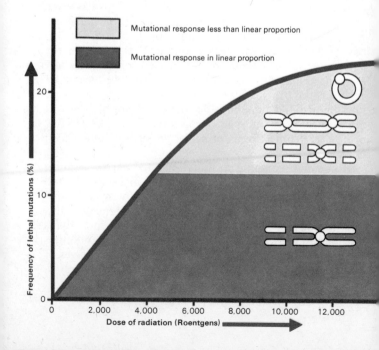

carrying it. When we consider inherited change we are talking about mutations among the cells which form gametes, for we can only detect this sort of variation in the offspring.

## Induced mutation

Although it was known early in the twentieth century that X-rays broke chromosomes it was not until 1927 that Muller first reported X-ray induced mutations as inherited changes. Using other forms of radiation such as ultra-violet light it was found that the most effective ultra-violet wave length for increasing the mutation rate was 260 nanometers, which is the wavelength of maximum absorption of nucleic acids, thus implicating DNA in the mechanism of mutation. Irradiated cells frequently show fragmented chromosomes, and the number of breaks shown by the chromosomes is related to the dose of radiation. Moreover, since there are mechanisms whereby chromosomes can repair themselves when they break, the fragments produced by irradiation can occasionally rejoin as inversions or translocations. This is sometimes enough to produce a change in phenotype. If the fragments are lost then there is an actual loss of genetic material which sometimes changes the phenotype sufficiently to produce death of the cell or individual carrying such a deletion.

Two main sources of radiation which can affect man are background radiation from the decay of radioactive elements in rocks and from cosmic rays, and medical X-rays. In addition there is the radiation hazard of atomic fall-out. Now if most of the genetically unfit individuals in the population were there as a direct result of mutation, then anything which increased the mutation rate of the population would have dire genetic consequences. However, it is possible that the genetically unfit individuals are present in the main as a result of their parents' being heterozygous and the genes at a given locus being part of a balanced polymorphism (page 120). The mutational possibility is called the *mutational load* of the population and is the price an organism has to pay for the privilege of being able to evolve by mutationally produced variation. The second possibility is called the *segregational load* of the population and is the price organisms pay for the Mendelian consequences of mitosis and meiosis. Studies have been carried out to see which load is the greater. It looks as though the mutational load is a comparatively small fraction of the total genetic load. If this is the case the gloomy predictions of early campaigners for nuclear disarmament may not need to be taken quite so seriously as they would have had us believe. Nevertheless, there is still a real burden in terms of human suffering which civilized man cannot ignore.

Chemical mutagens have proved to be exceedingly valuable tools in elucidating the coding structure of DNA. Ever since 1940 when Charlotte Auerbach demonstrated mutagenesis by mustard gas and related compounds in

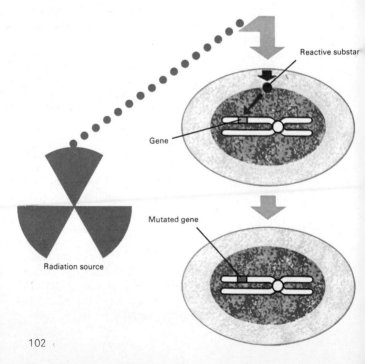

Radiation may induce mutation by forming a short-lived, highly reactive substance, such as a peroxide, which acts upon the DNA of the gene.

*Drosophila*, many compounds normally thought of as non-toxic have been found to be chemical mutagens under specific conditions. Many of the compounds studied have been shown to have a direct chemical action on the bases in DNA, which affects the bonding between the two DNA strands. Some of them are analogues of the bases and, tending to replace the correct base, alter the pairing properties at that region of the DNA thus forcing the 'wrong' base into the coding sequence during DNA synthesis and replication.

Earlier in this section, we mentioned the possible connection between mutation and cancer. This was not meant to imply that all cancer is produced by mutation. Indeed we know that this is not so: there are many agents which cause cancer which do not produce mutation. That mutagens tend to cause cancer is, however, likely to be true. By looking for chromosome breakage as an indicator of the mutagenic action of a chemical we can get an easy indication of whether that compound is a possible carcinogen. This is why scientists have looked at the chromosome breakage action of some of the drugs which are becoming part of our social scene. Young drug addicts may not be persuaded about the possibility of their passing on drug induced mutation to their children. It is, in fact, difficult to see how using LSD could affect the germ cells. However, there is certainly evidence that LSD breaks human chromosomes and this might implicate it in the causation of cancer.

We must also point out that it is now known that there is a very definite dose-response relationship between X-rays and point mutation. Even very low doses do cause point mutation. This is why X-ray technicians and radiographers always wear lead aprons to protect their gonads. We should also be careful not to indulge in unnecessary exposure to X-rays such as used to occur in seeing whether shoes fitted properly by using an X-ray machine in shops. In practice the risks involved in *not* being X-rayed for medical reasons are much more severe than the genetic hazard. This is why medical X-rays are allowed, and why the general public need not be unduly worried about them from the genetic point of view.

# POPULATION GENETICS

## Gene frequency

Let us consider the distribution of the genotypes of a hypothetical recessive condition on two islands.

|          | $AA$ | $Aa$ | $aa$ |
|----------|------|------|------|
| Island 1 | 16%  | 48%  | 36%  |
| Island 2 | 49%  | 42%  | 9%   |

The incidence of people with the condition differs on the two islands. Thirty-six per cent of the population of island 1 will suffer from it, whereas only 9 per cent will be affected on island 2. If we assume that only 100 people occupy each

Assuming random mating, genotype frequencies within a population are derivable from gene frequencies.

| 2nd chromosome ↓ \ 1st chromosome → | Gene A (p) | Gene a (q) | Total |
|---|---|---|---|
| Gene A (p) | AA ($p^2$) | Aa ($pq$) | $p^2 + pq$ |
| Gene a (q) | Aa ($pq$) | aa ($q^2$) | $pq + q^2$ |
| Total | $p^2 + pq$ | $pq + q^2$ | $p^2 + 2pq + q^2$ ($=1$) |

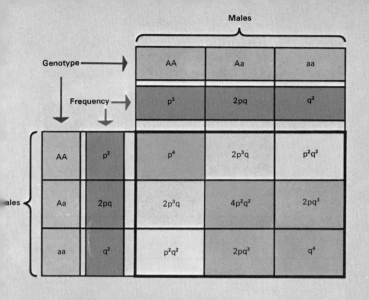

The frequencies with which certain matings will occur can be deduced from the genotype frequencies.

island we can trace the difference to a difference of the frequency of the two genes $A$ and $a$ on each island. We can arrive at the frequency by simply counting the genes present.

Each individual is diploid, therefore the 16 $AA$ individuals on island 1 contain $16 \times 2 = 32$ $A$ genes. The 48 $Aa$ individuals each contain one $A$ and one $a$ gene. The 36 $aa$ individuals carry no $A$ gene. Therefore of the possible $2 \times 100$ loci in those 100 diploid individuals $32 + 48 = 80$ of them are occupied by $A$ genes. The frequency of the $A$ genes is therefore $80/200 = 0 \cdot 4$. Similarly, the frequency of the $a$ genes on island 1 will be found to be $120/200 = 0 \cdot 6$. Whereas for island 2 the frequency of the $A$ gene will be $0 \cdot 7$ and that for the $a$ gene will be $0 \cdot 3$.

Thus the difference in *genotype frequencies* is basically due to a difference in *gene frequencies*. Given a population in which the genotype frequencies are known we can calculate the gene frequencies in that population. Let p be the frequency of the *A* gene and q the frequency of the *a* gene. Since no other genes will be assumed to occupy the locus concerned, $(p+q)=1$. Therefore for a given locus the total likelihood of its being occupied by an *A* or an *a* gene (that is of being occupied at all) will be $(p+q)=1$.

Knowing the frequencies of the different matings and the genetic outcome of each mating, the genotype frequencies in the next generation can be predicted. In the absence of disturbing factors these frequencies remain constant from one generation to the next (Hardy-Weinberg Law).

| Mating | | Frequency | | Frequency of offspring | | |
|---|---|---|---|---|---|---|
| | | | | AA | Aa | aa |
| AA × AA | ▶ | $p^4$ | ▶ | $p^4$ | | |
| AA × Aa | ▶ | $4p^3q$ | ▶ | $2p^3q$ | $2p^3q$ | |
| AA × aa | ▶ | $2p^2q^2$ | ▶ | | $2p^2q^2$ | |
| Aa × Aa | ▶ | $4p^2q^2$ | ▶ | $p^2q^2$ | $2p^2q^2$ | $p^2q^2$ |
| Aa × aa | ▶ | $4pq^3$ | ▶ | | $2pq^3$ | $2pq^3$ |
| aa × aa | ▶ | $q^4$ | ▶ | | | $q^4$ |
| | | | | $p^2$ | $2pq$ | $q^2$ |
| | | | | Total | | |

$$AA = p^4 + 2p^3q + p^2q^2 = p^2\,(p^2 + 2pq + q^2)$$

$$Aa = 2p^3q + 4p^2q^2 + 2pq^3 = 2pq\,(p^2 + 2pq + q^2)$$

$$aa = q^4 + 2pq^3 + p^2q^2 = q^2\,(p^2 + 2pq + q^2)$$

(Note that these two sentences are saying the same thing, but one way has turned the frequencies into a likelihood, or a probability.) Since individuals are diploid we will be concerned always with the likelihood of getting two genes together in one individual. If the chance of getting one *A* gene is p the chance of getting the two *A* genes together in one individual is p². The likelihoods of the various combinations are:

$AA$ — $p^2$
$Aa$ — $2pq$
$aa$ — $q^2$
Total = $p^2 + 2pq + q^2$
= $(p+q)^2 = 1^2$

Therefore, if p for *A* on island 1 is 0·4, and q for *a* on island 1 is 0·6, then the distribution of the genotypes – $p^2 + 2pq + q^2$ – becomes $(0·4)^2 + 2(0·4)(0·6) + (0·6)^2 = 0·16 + 0·48 + 0·36$ which is the same as the result we started with.

Thus working from a gene frequency we can predict the

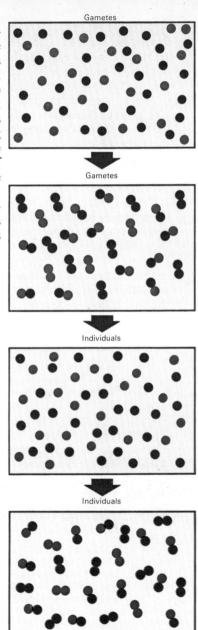

In a population to which the Hardy-Weinberg Law applies the gene frequencies remain constant through successive generations gametes and individuals.

genotype frequency as well as working in the reverse direction. We can calculate the gene frequency for a recessive condition in another way. If it is known that a disease is inherited as a recessive, and has an incidence of one in 10,000 of the population, then each affected individual is a homozygous recessive, *aa*, whose frequency is $q^2$. The gene frequency for that condition is 0·01 (=q) since $0·01^2 = 0·0001$. And since $p + q = 1$ we can see that the incidence of carriers of that gene will be $2 \times 0·99 \times 0·01$ ($= 2pq$) $\simeq 0·02$, or one in 50. A knowledge of the incidence of carriers is useful to the doctor when he is counselling a person who is the carrier for a recessive gene. If that person marries another heterozygote then one in four of their children will be affected. People who suspect that they are carriers (from a knowledge of their pedigree) want to know their risk of having affected children. The overall risk is not one in four, because if they married a homozygous normal individual then none of their children would be affected. To give them accurate information the doctor must give them the risk of marrying a heterozygote as well as the one in four risk of what will happen if they do. Therefore, in the condition we have been considering their risk of having affected children if they marry an apparently normal person is 1/50 (the risk of marrying a heterozygote) × 1/4 (the risk of affected children from an $Aa \times Aa$ mating) which is 1/200, a far better prognosis for the carrier seeking advice than 1/4.

## The Hardy-Weinberg Law

The next question one can ask is what happens to the gene frequency in the next generation? If we take the trouble to calculate the gene frequency in one generation, what use will this information be in thirty years time? Let us assume again we have a population of individuals where the gene

In a small population random fluctuation in numbers may have a disturbing effect on the genetic structure of the population. The obliteration of family c, with a predominance of certain genes when compared with the mean for the population, not only restricts the size of the gene pool but alters the gene frequencies within it.

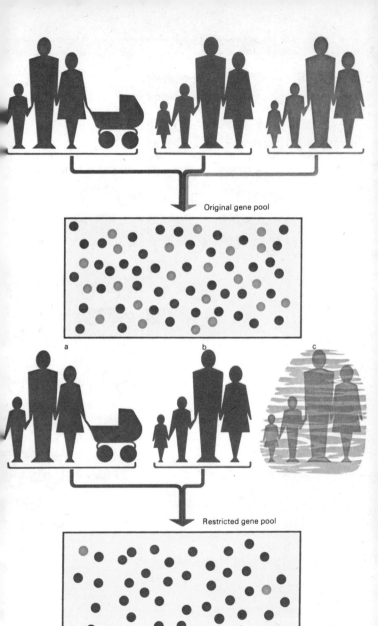

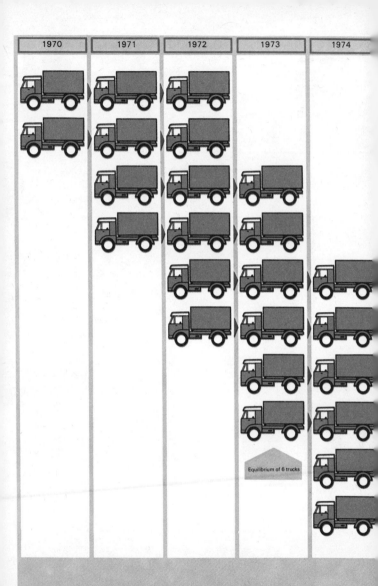

frequency of *A* gene is p and that for the *a* gene is q. The gene and genotype frequencies are shown in the illustration. What sorts of matings will occur in this population and how frequent will they be?

Let us assume that there is no difference in frequency of the genotypes in the two sexes. Let us also assume that mating is at random with respect to the *A* and *a* genes. This is not an unreasonable assumption for many genes. You do not bother about whether your prospective wife or husband is A, B, AB or O when you propose marriage to them. Neither do you enquire whether they have one particular type of enzyme in their serum. On the other hand, matings are not at random for things like height or intelligence, or even blindness or deafness.

The frequencies of matings, and the results of these matings are shown in the table. From totalling the offspring it can be seen that in the next generation the genotypes are arranged in the same frequency as they were among the parents. This fact was found independently by the Cambridge mathematician, G. H. Hardy, and the German geneticist, W. Weinberg, in 1908, and is now called the *Hardy-Weinberg Law*. A population for which this applied is said to be in *Hardy-Weinberg equilibrium*. Another way of saying what this means is to point out that if a population is in Hardy-Weinberg equilibrium or *genetic* equilibrium for a particular gene, the frequency of that gene in the population remains constant from one generation to another.

If this were to hold over thousands of years there would be no evolution, for evolution implies that there are new genes and therefore new genotypes arising in a population. We will discuss this later. For the moment, however, the genetic stability of populations enables us to pool data from several generations to test genetic hypotheses. For example, if we believe that a gene is recessive, and the population is in equilibrium, we don't have to rely on single families for testing for the Mendelian ratios. We can pool

Muller's trucking company buys two new trucks each year and keeps a truck in service for three years.

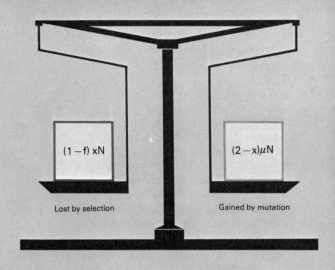

Calculation of mutation rate for a dominant mutant gene by the indirect method. The mutation of a particular gene is balanced by its loss through selection. Where N is the number of individuals in the population, x the proportion of affected individuals, and f the 'fitness' of affected individuals, then the mutation rate $\mu$, is approximately equal to $\frac{1}{2} x(1-f)$.

the data from all the families (whatever their generation) and look for the ratio we expect from the types of mating. This gives us much larger numbers for the statistical testing of the mode of inheritance of a particular condition.

There are four main factors which might be expected to upset a Hardy-Weinberg equilibrium.

## Mutation

When genes change from one gene to another, say $A$ gene into $a$ gene, or $a$ gene into $A$ gene, then we say that the gene has mutated. The new gene is a mutant gene. All alleles are mutants from a single, original gene occupying that locus. It is by the formation of mutant genes that variety is introduced into the gene pool of a particular species.

## Selection

When genes mutate, the individuals carrying the mutant gene are tested by the environment for their ability to survive. More particularly, survival implies that they can transmit that mutant gene to their offspring. There will be several genotypes for each locus, and the Darwinian theory of natural selection implies that the different genotypes may contribute differentially to the gene pool of the succeeding generations. If, say, homozygous recessive individuals have, on average, one child as compared with the heterozygous individuals or homozygous dominant individuals who have three children, then the dominant gene stands a greater chance of being passed on than the recessive gene. This enables biologists to give a precise definition of *fitness*, for comparing the number of offspring from individuals of two different genotypes, the relative fitness described in these terms can be quantified. Further, if the fitness of the fittest genotypes is given as 1, any reduction from this must measure the *intensity of selection* against individuals of the less fit genotype. In biological terms, the fitness (or as it is sometimes called, the *reproductive fitness*) of an individual can be influenced by a number of factors: fecundity, intensity of sex drive, mechanisms of pollination in plants, etc. These are the agencies through which natural selection can act on individuals.

## Immigration and emigration

When considering the population which may be in genetic equilibrium we must consider it either as a closed population in which there are no movements of individuals in or out, or we must assume that any immigration or emigration does not affect differentially any of the genotypes. An invasion of individuals from a population with a high blood group O frequency into a population for which that frequency was low would be expected to upset the genetic equilibrium of the latter population.

## Size of population

The size of the population must be large enough for random fluctuations in the individuals to count for very little. If in an isolated village in which people tended to intermarry and live near one another a flood occurred which killed families at one end of the village, this could completely upset any genetic equilibrium existing because the population was so small. An opposite situation occurs where a few individuals from a large population become the founders of a new one. This happened during the colonization of islands such as Tristan de Cunha. We do not expect the 'founders' to be a typical sample of the population, in terms of their gene frequency, from which they came, and therefore we would not expect, and do not find, that the gene frequency of the present islands is the same as the present parent population in other parts of the world. Such *founder effects* are known elsewhere. One of the best documented is that of the gene for *porphyria variegata* in South Africa. People suffering

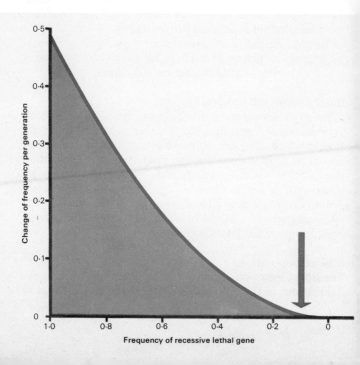

The rate of decrease in the frequency of a recessive lethal gene depends upon the gene frequency, becoming less as the frequency falls.

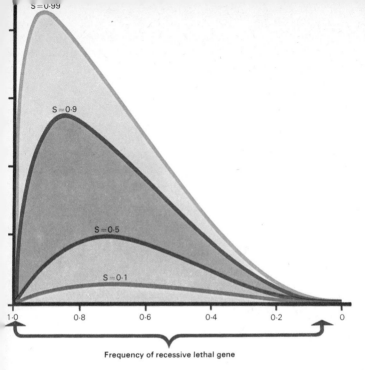

Frequency of recessive lethal gene

The rate of decrease in the frequency of a detrimental recessive gene depends upon the gene frequency and upon the selective pressure (s) against it.

from this condition suffer acute attacks after receiving barbiturates and sulphonamide drugs. Incidentally, this shows how the administration of drugs can bring out human variation that was previously unsuspected, and has led to the branch of genetics known as *pharmacogenetics*. Dean, in 1963, traced about 8,000 people in South Africa with porphyria variegata to Gerrit Jansz, who emigrated from Holland to the Cape in 1685. He was unmarried when he emigrated. The Dutch East India Company sent eight orphan girls out to the Cape from Rotterdam in 1688 to provide wives for the settlers. One, Ariaantji Jacobs, married Gerrit Jansz. Dean was able to trace every single family group back to this mating. The condition is also of interest because Macalpine and Hunter have suggested that porphyria was the condition responsible for the madness of George III.

## Selection and mutant genes

Now, although these factors may upset a genetic equilibrium there is evidence that what they may do is to establish a new one. In the illustration of Muller's Trucking Company, if we allow the time which the truck is kept to represent the statistical length of life of a new mutant gene, and purchase of trucks each year to represent the rate at which the new genes are formed by mutation, we see that just as the trucking company reached an equilibrium of six trucks, so the gene frequency would reach an equilibrium value. In practice mutation and selection act together. As new genes are produced by mutation so individuals carrying them may be selected against, or in rare situations selected *for*. The American geneticist Li has produced a simple model which has shown how these two forces ultimately reach an equilibrium.

Let A give B ½ his money.
Let B give A ¼ his money.

In a population where the alleles *R* and *r* occur with equal frequency, selection against the heterozygote *Rr* does not alter the gene frequencies.

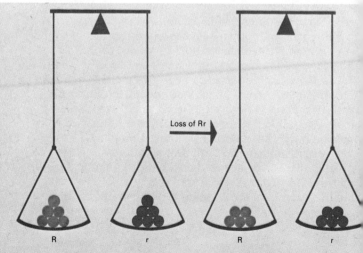

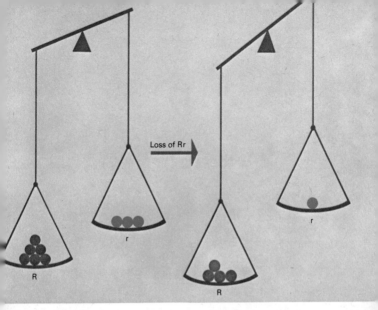

Selection against the heterozygote in a population where the frequencies of the two alleles are unequal shifts the frequencies to greater inequality.

A starts with £20
B starts with £280

|   | A | B |
|---|---|---|
| 1. | 20 | 280 |
| 2. | 80 | 220 |
| 3. |  | 205 |
| 4. | 98·75 | 201·75 |
| Tending to equilibrium at – | 100 | 200 |

People have from time to time suggested that in situations where a gene is very deleterious we might sterilize those who are affected by possessing it. This amounts to a selection against those who are homozygous for a deleterious recessive gene. It can be shown that for a gene whose frequency was about one in 50, giving an incidence of four per 10,000, it would take 50 generations (about 1,500

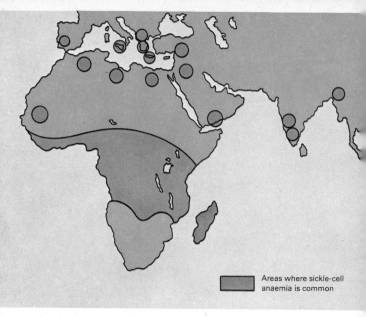

The distribution of sickle-cell anaemia.

years) to halve the gene frequency and this condition would still give an incidence of one per 10,000. Also this ignores recurrent mutation. Thus, as a practical genetic measure sterilization is not really a feasible proposition even if it were morally acceptable.

Two interesting ways in which selection can work are worth discussing.

*Selection against a heterozygote:* one situation in which this applies concerns the genes responsible for a woman being Rhesus negative. Since she will put out antibodies to any Rhesus positive child – and the child will be heterozygous – in the past this amounted to selection against a heterozygote, for the children, especially third and fourth children, died from haemolytic disease of the newborn. From the illustration it can be seen that this type of selection only leads to a stable equilibrium when the genes concerned are equally frequent. In other situations, whichever gene is less frequent is eliminated. Since the proportion of Rhesus negative women in Europe is about 14 per cent, this value is extremely high for a 'gene' which ought to be eliminated. No very satisfactory explanation of this apparent contradiction has been forthcoming. Unfortunately (in one sense), with the advent of exchange transfusion for the children and the possible immunization of the mothers by Clarke's method, we are unlikely to find the answer, for now we have altered the selection process itself.

*Selection for a heterozygote:* if the heterozygote is favoured as compared with either homozygote then we end up with most matings being between heterozygotes. This in turn means that individuals of both homozygous genotypes will be born as they segregate from such matings. However, they will occur in quite appreciable frequencies, in spite of selection acting against them. The best worked-out case of this is in the sickle-cell gene.

Selection for the heterozygote: sickle-cell trait confers a degree of resistance to malaria.

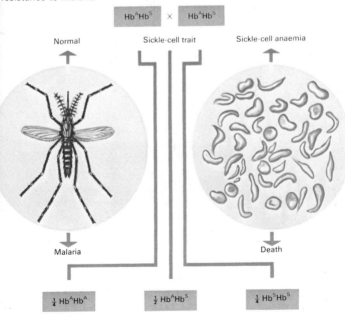

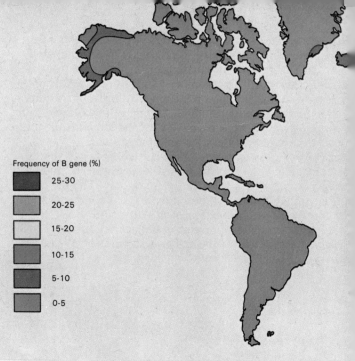

The distribution of blood group B may be linked with the prevalence of smallpox.

Hb$^A$ Hb$^A$  Hb$^A$ Hb$^S$  Hb$^S$ Hb$^S$
Normal  Sickle-cell trait  Sickle-cell anaemia

In parts of Africa around the equator the frequency of the Hb$^S$ gene is 0·05–0·2 giving an incidence of sickle-cell anaemia of as high as 4 per cent, and yet sickle-cell anaemia is a disease which is extremely serious with a very low fitness. Allison was able to show that people who had the benign sickle-cell trait contracted malignant tertian malaria far less often than *normal* people in these areas. Thus malaria was a major selective agent *against* the normal and *for* the heterozygous individuals who had sickle-cell trait. This meant that the population was kept in genetic equilibrium in the different forms. E. B. Ford at Oxford gave the term *balanced polymorphism* to such a situation. He laid

emphasis on the fact that in such a condition the frequency of the least frequent gene was higher than could be accounted for by recurrent mutation alone. As workers have studied normal human variation, especially in terms of biochemical variation, they have found such variation to exist as balanced polymorphisms. We now suspect that we are all heterozygous at many of our loci. Studies on animals and plants have shown how advantageous this can be for each species. Crossing inbred strains to form heterozygotes often produces offspring which are more vigorous than their parents *(hybrid vigour)*. If a dominant phenotype is presented to natural selection as a heterozygote a hidden variability is preserved in terms of the recessive genes should selective forces change. The homozygous recessive offspring *may* turn out to be better adapted if the environment does change. Much has been written on this and readers are referred to page 156 for an introduction to the extensive literature. Suffice it to say that one of the major problems in human genetics is to understand the selective forces which maintain many of our balanced polymorphisms. Human geneticists are looking for the 'diseases', like malaria, which maintain each system.

## Variability and genetic change

If we consider a population of organisms as a collection of genes, sometimes called a *gene pool*, we can now see the basic requirements of that population from an evolutionary standpoint. Firstly the genes must produce phenotypic characteristics which make the individuals in the population well adapted. Secondly, since the environment may change so that adaptation becomes less satisfactory, the gene pool must possess the means of making the population variable so that natural selection can favour some of the variants. These two requirements look mutually contradictory. A population could be well adapted by selecting out disadvantageous mutant genes and being completely homo-

Light and dark variants of the same species of moth are clearly adapted to differently coloured backgrounds. Pollution has brought about a marked increase in the proportion of dark moths in affected areas.

zygous. However, if this were to happen there would be no possibility of variation other than by occasional mutation and in the event of a new type of selection the right mutant might not be available at the right time. The solution to the dilemma for the population would be to let some genes behave as dominants and recessives. A population of heterozygotes having dominant genes, where the characteristic produced by the dominant gene was that of the best adapted homozygote, would itself be well adapted, *and* would contain recessive genes as part of its hidden variability. If, at most loci, the genes produced balanced polymorphisms, then when selective forces in the environment changed, the hidden variability would produce a large number of organisms sharing changed characteristics, which might have a higher survival chance.

These two needs, stability and variability, are still not sufficient for evolution and the formation of new species. The third requirement is that a change, environmental or internal (often cytogenetic), must occur whereby the original gene pool is split. This isolation of part of the gene pool – so that this part is not able to freely mingle back into the original gene pool – is what *species formation* is really about. We do not have space to consider the various isolating mechanisms which can occur.

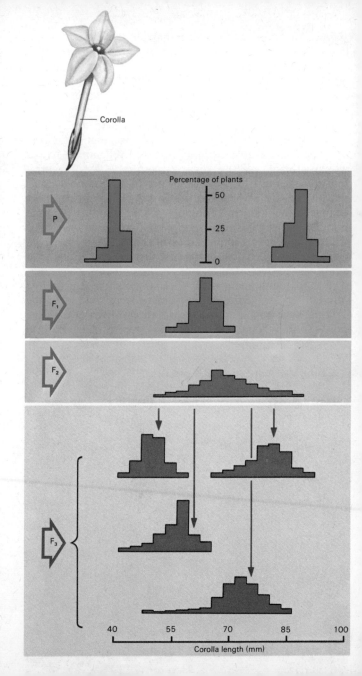

# MULTIFACTORIAL INHERITANCE

Mendel was lucky in that he studied the inheritance of characters which varied discontinuously in his population of peas. Galton was unlucky for he was led to a theory of blending inheritance because he studied characters such as height, which vary in a continuous fashion in the population. By the end of the nineteenth century when Mendel's work was rediscovered, attempts had to be made to explain the sort of results Galton had obtained on the basis of the Mendelian, non-blending, particular model.

Johannsen (1903), Nilsson-Ehle (1909) and East (1910) working on crop plants like beans, oats, maize and wheat in terms of the inheritance of bean weight, or the number of rows of grain in an ear of maize, were the first to demonstrate how it was possible to explain continuous variation in Mendelian terms. They suggested that if one assumed that many loci were involved this could be done. Since the involvement of many loci must imply that many alleles or factors could occupy those loci, they called the concept *multifactorial inheritance*. Moreover, the environment was one of the factors which could produce variation. Realising this, Johannsen invented the terms *genotype* and *phenotype* to distinguish between the *hereditary constitution* of the individual and the *appearance* of the individual which must be the result of the interaction between genotype and environment.

## Continuous variation

If we look at a very simple model we may understand how, in principle, several genes could begin to give a picture of continuous variation. Suppose that at one locus there are three alleles concerned with height. If we assume that each allele contributes so much to the height of an individual

The inheritance of corolla length in *Nicotiana*. The means of the first and second generations are intermediate between those of the parents. The means of the four third generation families are correlated with the corolla lengths in the second generation plants from which they came.

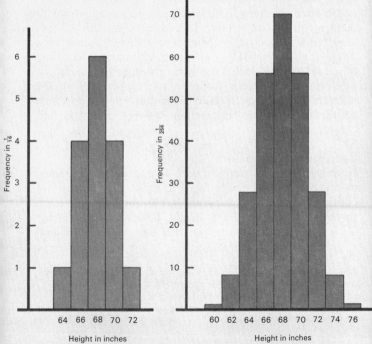

carrying it, and that each allele's effect can be added to another's contribution, then one allele, $H$, could produce the average height for the group of individuals, say 68 inches, while of the other two alleles, $h-$ and $h+$, $h-$ could reduce height by 2 inches and $h+$ increase height by 2 inches. We would then have six types of individual:

| | |
|---|---|
| $h-h-$ | 64 inches |
| $h-H$ | 66 inches |
| $h-h+$ | 68 inches |
| $HH$ | 68 inches |
| $Hh+$ | 70 inches |
| $h+h+$ | 72 inches |

We now have a discontinuous classification by genotype of different heights. In order to see how the genotypes are distributed in the population, we must consider the frequency of each gene. With this in mind we see that we can draw a histogram of the distribution of height in such a population. We must note that the shape of the histogram actually depends on the frequency of each gene in the population.

If we now added another locus with other genes $T$ (average 68), $t-$ (reduction by 2 inches) and $t+$ (increase by 2 inches), and allow independent assortment of these genes with respect to the locus for the $H$ genes, again we get histograms for the distribution of height. In fact, as we add further loci the frequency distribution of heights becomes smoother in shape. Were we to allow environmental variation at each height, we can see that the histogram would become the smooth curve shown in the population by the continuous variable.

Not every form of variation is inherited, whether it is discontinuous or continuous. We get a hint that a particular variant might be inherited when that variant is commoner in relatives of people with it than in the general population.

*(Left)* The distribution of height produced by three alleles at a single gene locus. *(Right)* The distribution of height produced by three alleles at each of two loci.

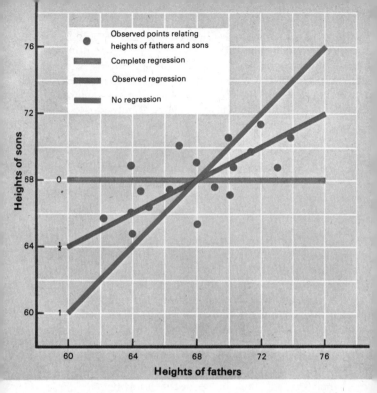

The regression of the mean of sons' heights on fathers' heights

That is to say, there is a family incidence of the variation. If you think about the Mendelian pattern of inheritance this certainly holds. It holds equally for variations which are continuous. Tall people's relatives are taller than the average for the general population. However, we must be clear about what we mean by tall. We must refer to the height of an individual as compared to the average of the population in order to get some measure of what we mean by tallness. Thus, even when we consider an individual's height we have to have half an eye on the population average height in order to assess how much he differs from the average. What we said about the genes at a given locus reducing or increasing height was referring to their contribution towards that individual's difference from the average height.

If we were to take a number of men whose height was about 74 inches and mate them to women whose height was about 68 inches, then the offspring from these matings would vary, but their *average* height would be about 71 inches. In other words the average height of the offspring would be half way between the parental average heights. Statisticians say that the offspring height *regresses* to half way between the male and the female population means. A line could be drawn relating the heights of male parent and offspring. This line is called a *regression line*.

Two points can be made. Firstly, we see that the parent/offspring relationship is now not in the terms of individuals but in terms of groups of parents and groups of offspring, because we tend to talk about average values. This means that ways of handling breeding programmes in studies of

The genetic resemblance between parents and offspring is ½ since each child inherits half its genes from one parent.

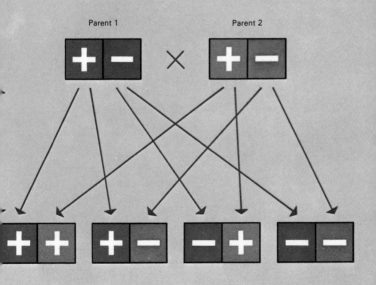

continuous variation rely on the statistical treatment of data which is beyond the scope of this book. Secondly, we can now ask why, and under what conditions, offspring heights regress to half-way between the average heights of the two parental groups, that is towards the mean for the total population. To understand this, we must look at the degree of resemblance in genetic terms between various groups of relatives.

*Parents/offspring:* a mother or father hands exactly half of their genes to a child in gamete formation. In the case of a continuously varying trait which is almost completely genetically determined such as height, the regression we spoke of reflects this contribution as a shift of mean offspring height half-way towards the population average. The genetic resemblance between parents and offspring is one half.

The genetic resemblance between siblings is ½ since *on average* half their genes will be common.

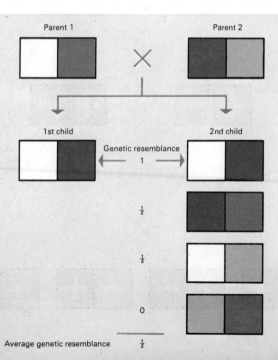

*Identical twins:* these are formed as the result of a mitotic division of the zygote. They are more accurately called *monozygotic twins.* Since they come from the same zygote they should be identical genetically. We then say that this genetic resemblance is one. In other words we should expect such twins to be the same height and any deviation from this for a twin-pair must be entirely due to environmental factors.

*Siblings:* if one considers two parents containing genes A and B, their offspring could be AA, AB or BB. Then looking at a pair of these offspring we could have: (a) first child AA, second child BB; (b) first child AA, second child AB; (c) first child AA, second child AA; as well as other combinations. In the first case (a) the siblings are different. In (b) they are similar for A, but not for B. In (c) they are identical. Considered over a large number of pairs of siblings and over many loci there is a likelihood that siblings will have half their genes in common. This is a purely statistical half as compared with the mechanical one in two for parent and offspring. Their genetic resemblance is again one half.

*First cousins:* the likelihood of cousins having genes in common is the product of the above resemblances:

Offspring/Parent ($\frac{1}{2}$) × Sibling pair ($\frac{1}{2}$) × Parent/Offspring ($\frac{1}{2}$) = Cousins (1/8)

## Heritability

Now it is possible to use the mathematical correlation between relatives to get a figure in any given case for actual genetic resemblance. As we have seen, when this is done for some human characters we find:

|  | Correlation coefficient | |
| --- | --- | --- |
|  | Parent/Offspring | Full siblings |
| Stature | 0·51 | 0·53 |
| Span | 0·45 | 0·54 |
| Intelligence | 0·49 | 0·49 |

The interpretation of these figures is that *if* any of these characteristics were completely genetic we would expect on genetic grounds the figures to be 0·5 for the correlation coefficient. Without being too precise, and since the figures are roughly what we expect, the discrepancy between observed and expected values gives an indication of the amount by which genetic factors contribute to the variation of the character. Looking at the figures one could argue that most of these characters are almost completely genetically determined. However, anything which would contribute to resemblances between individuals would also be involved. For example, marriages generally occur between people of similar intelligence. Indeed the correlation coefficient between husband and wife for intelligence is 0·68 (correlation coefficients go from $+1$ to $-1$), yet husbands and wives are, in most societies, never closer than cousins.

Therefore, in order to make more satisfactory estimates of the genetic contribution, known as the *heritability* of a condition, the geneticist has to seek out special breeding situations where other aspects of the genetic contribution as well as the environmental contribution can be evaluated. Indeed, it must be stressed that the heritability only applies to the particular population in which it was measured. An example of special breeding situations is found in the study of twins. Comparison of identical twins reared together and reared apart enables us to get a measure of environmental factors.

Let us summarize where we have got to. The *heritability* of a character tells us the magnitude of the genetic influences in the production of that character's variation from a particular population average. The theory behind this is that many loci, each with genes which contribute additively to the variation, are involved. Then by segregation at each locus and between loci we get the overall picture of continuous variation.

So far we have considered phenotypes which are obviously describable in terms of continuous variation. It so happens that the commonest human congenital conditions

Random segregation of a number of genes influencing a single character allows the possibility of considerable variation between parents and offspring.

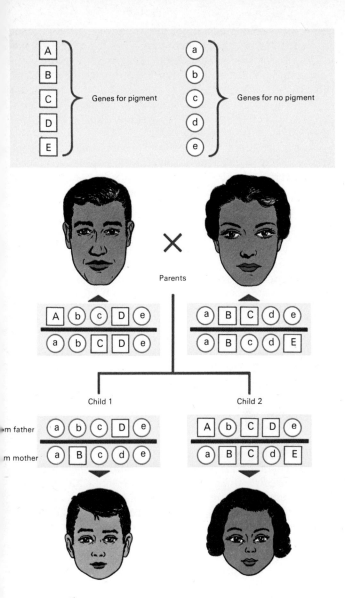

appear as variants which are discontinuous in the population and yet when their genetics are investigated they do not segregate in a Mendelian manner. Such conditions are clubfoot, cleft lip and palate, pyloric stenosis, certain brain and spinal cord conditions such as anencephally and spina bifida, congenital heart lesions, etc. They are common, which means that they have an incidence greater than one in 1,000.

If one examines the incidence of the condition in the relatives, it is higher than in the general population, but as can be seen in the table the incidence falls away dramatically as the degree of relationship, and the genetic resemblance, becomes more remote. These conditions have puzzled geneticists for their inheritance is difficult to explain. Doctors have had to rely on risk figures for counselling parents with an affected child. It is instructive to think first about Mendelian explanations.

If the condition were recessive we would find an appreciably higher incidence among siblings of affected individuals than among their children. Affected siblings from a mating of normal parents would be part of a three to one ratio of normal to affected offspring. The frequency of affected children arising from affected parents would depend upon the probability of mating with either a heterozygote or another homozygous recessive. Assuming that the recessive gene frequency would be low this probability would be correspondingly far smaller than one in four. If the condition were dominant every affected child would have an affected parent and there would not be the sharp fall off in the incidence as we looked at the relatives.

The explanation is that there is a continuously distributed *genetic predisposition* with some threshold value above which individuals are at risk of showing the condition. Environmental factors may be expected to contribute to any predispositions. Curves have now been constructed so that from knowing the incidence in the population one can predict the risk to relatives. This is very useful in counselling, but we have no idea of what is meant by the genetic predisposition in terms of gene action.

We must now try to clarify what might be meant by

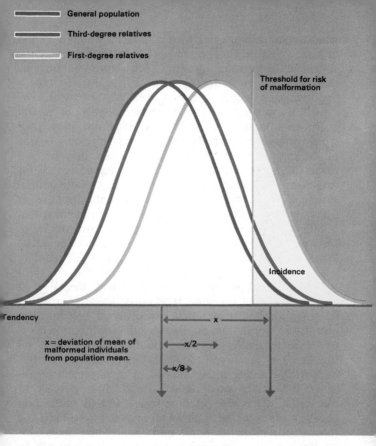

Threshold model for the incidence of disease in relatives of affected individuals assuming the multifactorial inheritance of a genetic predisposition.

multifactorial inheritance in terms of gene action. As we have described the Mendelian picture we have referred to a number of genes each contributing a quantity to the variation. We spoke of a gene adding 2 inches to the height. Some geneticists refer to these genes as *polygenes* and have suggested that they are associated with a particular part of the chromosomal material. I suspect that this is an erroneous view of how the genes work. It may well be that what

what we are calling a polygene is in fact a resultant action of some major genes. The concept has arisen because of the statistical treatment of our results, but we must not imagine that there are genes for '2 inches of height'. To take a hypothetical example, height may be partly regulated by growth hormone. If we found that some aspect of growth hormone formation or action was regulated by a gene which behaved in a Mendelian way, this gene may be one of the multifactorial system – and its effect may be '+2 inches of height'. In other words, we should search for single genes which may be involved in a multifactorial system. Thoday and his colleagues at Cambridge have, in a very elegant series of breeding experiments, shown that a difference in bristle number between two stocks of *Drosophila* is largely due to three single identifiable genes. Work and thinking along these lines is urgently needed to account for many interesting human conditions.

Multifactorial genetic methods have had their greatest impact in animal breeding. The introduction of artificial insemination has meant that experimental breeding can give vast information on the genetic contribution to a particular character because one bull can be used on thousands of cows spread among many different herds, each with differing environmental or management conditions. By various experimental crosses it is possible to assess the effects due to dominance and to interaction between genes in the multifactorial system. Perhaps the most important factor is the introduction of a method of quality control by *progeny testing*. A bull's semen can be tested on 500 cows and the offspring evaluated in terms of their variation round a mean milk yield, if this is the character which is being considered. From this the *breeding value* of the bull in terms of average gain of milk yield can be assessed. Farmers can expect a rise of so many gallons if that bull is used on herds subsequent to the progeny testing. Equally, bulls can be rejected if their performance on the progeny test is not satisfactory. In the period, which may be four to five years, over which progeny testing is carried out, the bull's semen can be collected and stored for possible eventual use. Since this requires recording yield as

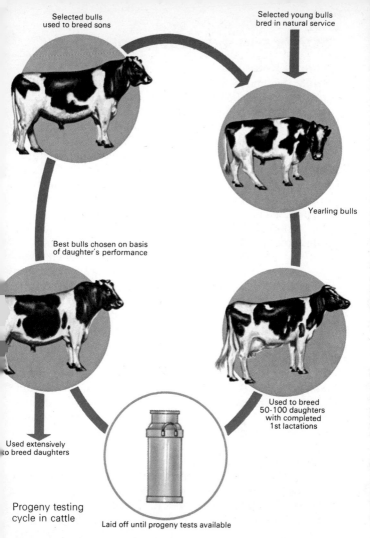

Progeny testing cycle in cattle

part of herd management, farmers can also become aware of any deleterious effects of inbreeding that can happen. When or if this does occur it is relatively easy to bring new genes into the herd by artificial insemination and at the same time know something about the performance of new genes.

# GENETICS IN THE SERVICE OF MANKIND

## Animal breeding

Before the time of Mendel animal breeds consisted of local subpopulations, which differed in various production characteristics. Local enthusiasts selected for practical characteristics, but this had little value. Mendel had little impact until 1910. People tried to use Mendelism for various characters such as fat content of milk and milk yield itself, but found that the inheritance of these characteristics did not behave in a Mendelian way. Most of the important practical needs involved continuously varying characteristics and this had to wait until the 1930s when Wright, Lush and Lerner in America were able to develop the theory of multifactorial genetics to apply to animal breeding. Although the theory allows us to consider the genetic resemblances between animals and to predict the effect of matings, it has not led to any new breeding programmes. Lerner attempted to develop new mutants in animals, but this also has not been successful. Irradiation of sperm led to the development of deleterious mutants and he was not able to utilize useful ones.

Recently Robertson showed that such developments as took place did so mainly for reasons of husbandry rather than genetics. The economics of poultry and pig breeding involving very large breeding units has been more effective in determining genetic development than the genetics itself. Artificial insemination has meant that the dairy industry can proceed without any interest being taken by many producers in cattle breeding as such. Recently the advantages of the situation have been realized by research centres. Artificial insemination gives information on a breeding unit of 50,000 animals, and it is possible to assess the daughters in a wide variety of environments. It is now known that cattle possess about ten blood group polymorphisms. Five of these are correlated with milk proteins and this information is useful in attempts to give controlled hybrid vigour.

Four different breeds of sheep

In poultry, crosses are made in the hope of producing heterozygotes with hybrid vigour. The males are selected on the basis of progeny testing. The broiler meat industry is an example. The females are selected for egg laying, so that in the end hybrids are made from the high-laying strains. On the other hand males are selected on the basis of growth rate up to eight weeks and carcass composition. These males are then crossed to hybrid females from high egg-laying strains to produce large numbers of broiler birds, with good growth rate and carcass qualities. Again the breeding units are by economic necessity very large. Because of the success in economic terms of the poultry breeders and the food suppliers the latter have at last made an impact on the pig and sheep industry. Indeed in some countries the latter industries have been taken over by the poultry breeders. The sheep industry has many breeds and yet the resulting animal tends to be a hybrid. Females are selected for maternal qualities, such as hardiness when carrying young and milk yield, so that a suitable hybrid female might be a cross between a mountain breed, such as the Welsh Blackface, and a Leicester, with a high milk yield. These females in turn are crossed with a male selected for growth rate and carcass qualities. In Australia the wool qualities will also be of paramount importance.

## Plant breeding

The plant breeders, especially those concerned with food crops, are faced with an enormous challenge in terms of world population. Geneticists have made major contributions to the solution of food shortage.

Borlock took a strain (Norin 10) of *Triticum*, a bread wheat, which was semi-dwarf and remained upright under high levels of fertilizer application, from Cambridge to Mexico. There he developed further strains which were day length neutral and had a short generation time in low

*(Upper)* The origin of compair wheat. *(Lower)* The five different grasses and wheats used. *L* to *r: Aegilops comosa; Triticum aestivum* (Chinese spring): 44 chromosome disomic addition of 2 M to Chinese spring; *Aegilops speltoides;* compair.

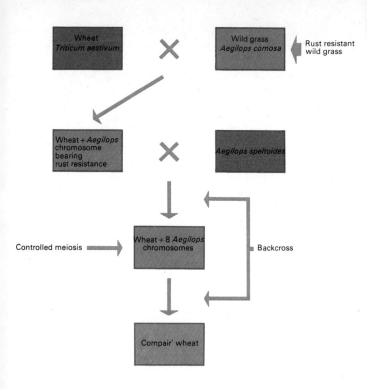

latitude regions. Between 1945 and 1956 Mexico was able to become self-sufficient and now exports liberally a wheat suitable for North Africa, India, Pakistan and similar countries. In 1968 Pakistan was able to become self-sufficient and, if present advances are maintained, India could be self-sufficient for this wheat by 1972. The short life cycle allows two or three crops per year.

Many of these advances depend upon developing disease resistance, for the new strains have to keep pace with mutations in disease producing agents such as fungi. The most ingenious development along these lines was carried out by Dr Ralph Riley of Cambridge. It was known that a form of rye was stem-rust resistant. Riley was able to get hybrid plants from wheat/rye crosses which contained both rye chromosomes and wheat chromosomes. Sears, of Missouri, discovered that on exposure to X-rays, the part of the rye chromosome which carried the gene for stem-rust resistance could be translocated onto a wheat chromosome. It was discovered that there was a gene on the long arm of one of the wheat chromosomes (5B) which restricted meiotic pairing to the wheat chromosomes. This inhibited the rye chromosomes from gamete formation. Riley was therefore left with gametes containing only wheat chromosomes but one of which carried the stem-rust resistance gene from rye – giving a new strain of disease resistant wheat. This is an elegant example of genetic engineering opening up the possibility of a vast new range of genetic variants.

Two hundred million people in Central and South America rely entirely on maize. However, maize is low in some amino acids such as lysine, methionine and threonine which are essential for a balanced human diet. The human disease *kwashiorkor* is the outcome of trying to live on an unbalanced diet lacking these essential amino acids. O. E. Nelson, at Purdue University, Indiana, developed in 1966 a cross between *Opaque 2* and *Floury 2*, two strains of maize which have a higher than normal amount of lysine content. Provided the yield of this cross does not drop, this should provide a better protein diet and it is hoped that the hybrid should be generally available for countries such as India in the early 1970s.

## Innovations in food production

A novel form of farming has arisen from the study of microorganisms such as bacteria. If one expresses the amount of protein that organisms can produce as pounds of protein per hundredweight of producing organisms per day, then a bullock produces 0·9 lbs and a soya bean 82 lbs, whereas yeast (a fungus) can produce 112,000 lbs. Workers have, therefore, looked at the possibility of increasing the world's supply of food by the use of microorganisms. These techniques are being developed in Japan. In 1967, an oil company built, in France, the first large scale plant in which yeast is grown in the hydrocarbon fraction from crude oil. There are aeration problems and cooling problems in the process for the microbial synthesis of protein generates large amounts of heat. Certainly this method of protein production is biologically feasible. Whether it is economically and gastronomically feasible only time will tell.

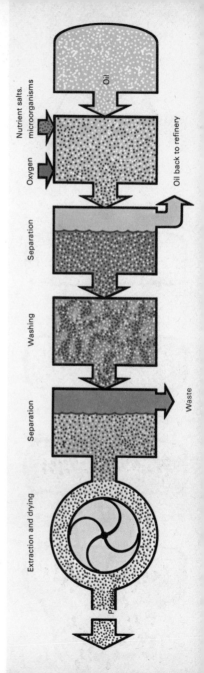

Yeast can be used to produce protein from crude oil.

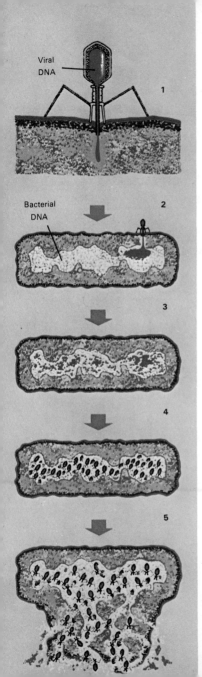

## EXPERIMENTAL ORGANISMS

### Phage

A bacteriophage, or phage, is a virus which lives on bacteria. It enters a bacterium, reproduces itself and eventually destroys its host *(lysis)*. When it does so a generation of new viruses then emerges. Each phage consists of two main parts: a central core of DNA surrounded by a protein coat. These two parts have separate functions. The ability to attach the phage to the bacterium resides in the protein coat, while the power to reproduce the phage lies in the DNA core. However, it has been shown that the DNA is also concerned in the formation of a new protein coat around daughter viruses.

When a phage attaches itself to a bacterium its DNA is

Stages in viral attack on a bacterium: (1) Viral sheath contracts driving the core through the cell wall. (2) Viral DNA enters the bacterium. (3) Bacterial DNA is disrupted and viral DNA is replicated. Structural proteins of virus are synthesized and assembled. (5) Cell finally bursts releasing new viruses.

put into the bacterium while the protein coat remains outside. Once inside the bacterium the phage DNA begins to make replicas of itself using as its raw materials, bases and other materials provided by the bacterium.

Phage is grown on thin films of sensitive bacteria which are cultured on agar plates. When lysis occurs the viruses reinfect other bacteria which in turn lyse and the process continues until a clear area is made visible on the agar plate. This area is called a *plaque*, and counting such plaques is the unit of observation of most phage genetic experiments.

The phages we have described are *virulent* phages and have been studied more intensively than the commoner *non-*

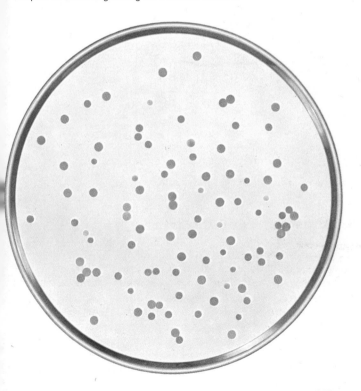

Plaques of bacteria growing on a culture medium

*virulent* or temperate phages. The latter do not kill the bacteria they enter. Bacteria which carry such pages are said to be *lysogenic* and do not liberate infectious phages. The phage in this state is called a *prophage*. There is evidence that at some stage the prophage is able to recombine into the bacterial chromosome, and that when it comes off the bacterium it is liable to take with it bacterial genes. This has turned out to be a vitally important happening for the study of the bacterial genotype.

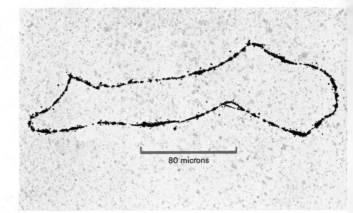

The ring structure of the bacterial chromosome is revealed by autoradiography of labelled DNA.

## Bacterium: *Escherichia coli*

Bacteria are able to pair and join (conjugate) with each other during their life cycle. Not all strains of bacteria will conjugate with each other. One strain is known as $F^-$, another strain is called $F^+$, and a third strain is called Hfr (high frequency of recombination). $F^-$ bacteria do not conjugate with each other but combine with $F^+$ and Hfr strains. When conjugation occurs a narrow cytoplasmic bridge joins the two bacteria. The attachment lasts about an hour then the conjugants separate and each divides to form a clone (daughter cells from one parent). Since the $F^-$

clone contains recombinant progeny, but not the Hfr clone, this suggests that material passes from the Hfr bacterium to the F⁻ one. It was found that the number of genes transferred from Hfr was directly proportional to the time elapsed before the conjugants were separated by a special technique. This method of interrupted mating has allowed the mapping of the bacterial chromosome.

The bacterium *E. coli* has been extensively used in these genetic studies. This is partly because it contains only one chromosome and partly because the replication of its DNA occupies the bulk of the cycle of cell division. Cairns was

Bacteria exchange genetic material through slender, tube-like appendages which are outgrowths of the cell wall.

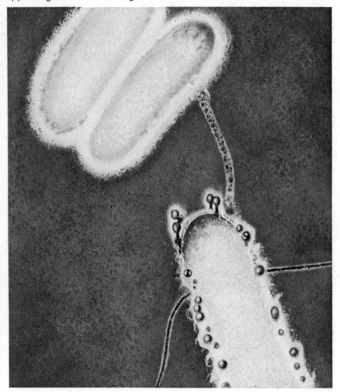

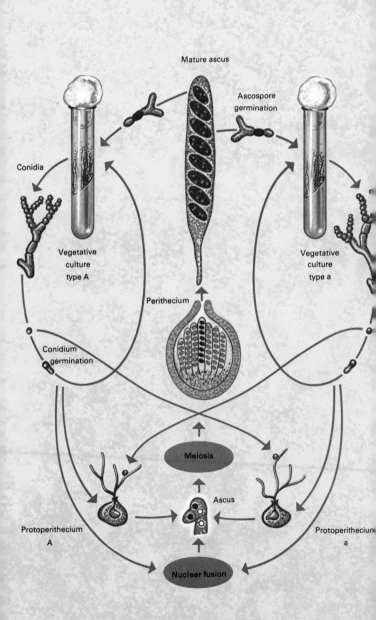

able, in 1963, to study the shape, appearance and replication of this one chromosome by using autoradiographic techniques. He found that *E. coli* had a chromosome which consisted of a single molecule of DNA roughly 1 mm long. When he studied the replication of this massive molecule he produced evidence to show that the bacterial chromosome was circular.

Extensive mapping has now been done of the chromosome of several bacteria, and the techniques have advanced so much that in 1969, Beckwith and his colleagues at Harvard were able to remove a single operon from a bacterial chromosome and initiate a new era of detailed biochemical studies of the behaviour of a single genetic unit.

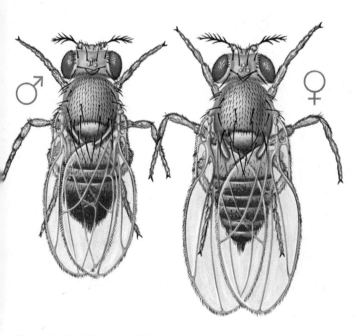

*(Opposite)* The life cycle of *Neurospora*.

*(Above) Drosophila melanogaster* has contributed much to experimental genetics.

## Neurospora

*Neurospora* is a bread mould. It has a brief life cycle and can be used in large quantities. Since it reproduces asexually, unlimited populations of a given genotype are readily available. One advantage it has is that it is very easy to grow on chemically defined media. This means that mutants for various nutritional requirements can easily be handled and can be selected for. At times in its life history meiosis occurs and the products of meiosis are released into small sacs called *asci* which can be easily studied. The normal habit of the mould is a set of fine branching tubes (hyphae) containing haploid nuclei. These haploid nuclei divide by mitosis and some find their way into small spores called *conidia*. The conidia can be blown about in the air, and when they settle onto a suitable substrate will germinate pushing out more haploid hyphae. One hypha may belong to one of two strains.

Gene controlled variation in *Drosophila:* (a) vestigial wing; (b) rudimentary wing; (c) twisted abdomen

When strains of opposing mating type meet they can fuse. In this case the haploid nuclei from each strain fuse to form a diploid zygote enclosed in the ascus. In the ascus meiosis with recombination occurs to form eight nuclei arranged in linear order up the narrow ascus as four pairs of spores, each pair represents one of the products of meiosis. This affords an exceptional opportunity to study the genetic consequences of meiosis, because each ascus contains the

Chromatograms of extracts of whole crushed *Drosophila* reveal the amounts of certain pigments known as pteridines present in different eye colour mutants. The pteridines are extracted with a suitable solvent and allowed to run on filter paper. The bands fluoresce in ultra-violet light.

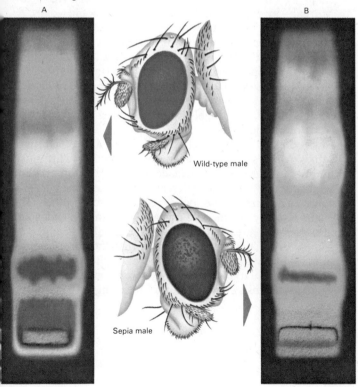

meiotic products of only one cell.

Occasionally the hyphae fuse without fusion of the nuclei. This leads to the formation of a *heterokaryon*, in which the haploid nuclei from each hypha can exist together. If one imagines that the hyphae (and therefore their nuclei) differ in respect of a nutritional requirement the formation of a heterokaryon provides the geneticist with a powerful tool with which the genetic lesions of the nutritional mutation can be analysed at the fine gene level.

## Drosophila

Much of the early support of the chromosome theory of heredity came from a study of *Drosophila melanogaster* – a small fly that one can see hovering over very ripe fruit. Morgan, in 1910, found that these flies could easily be reared in the laboratory and finding a white-eyed male mutant arrived at the theory of sex linkage. *Drosophila* are easily kept in small flasks, or traditionally milk bottles, where they reproduce with a generation time of about ten days. Mutants of eye colour, body colour, bristle shape and wing shape are easy to see under a good hand lens. *Drosophila* is therefore a highly suitable subject for laboratory study of classical genetics.

Moreover, *Drosophila* has only four pairs of chromosomes which are not difficult to distinguish. This small number of linkage groups made it possible to observe the segregation of alleles at different loci on the same chromosome and to map these loci in relation to each other. The chances of any two loci occurring on the same chromosome are obviously greater the smaller the number of chromosomes. Especially useful is the fact that the chromosomes of the salivary glands of *Drosophila* and other dipteran flies are extremely large (200 times the corresponding chromosomes at meiosis), and have very meaningful banding on

Larvae of *Drosophila* have salivary glands (1) containing large cells with large nuclei (2). The giant chromosomes (3) of these cells are easily visible once the cellular and nuclear membranes have been ruptured. The chromosomes show a specific, constant sequence of bands (4).

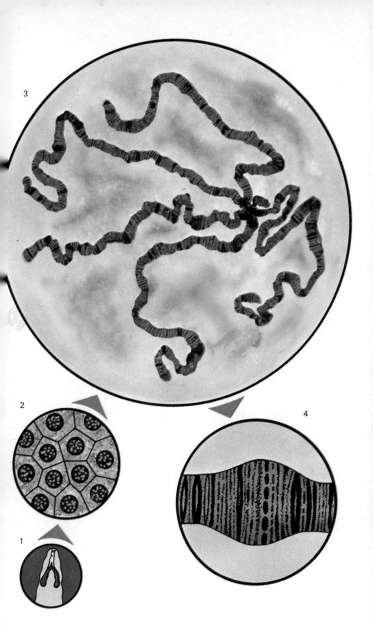

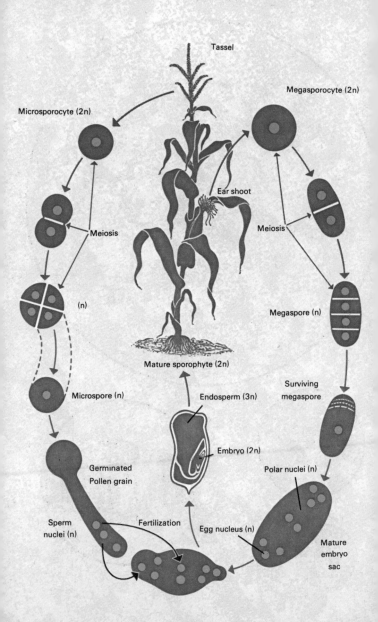

them visible under the microscope at all times in the life of the cell. By studying the bandings of the salivary gland chromosomes in *Drosophila* and noting that changes of the banding occurred with specific changes of phenotype of the fruit fly, advances were made in our knowledge of chromosome abberration. Duplication and deletion of chromosome material could be investigated and related to changes in phenotype. It was also possible to partially check the linkage map against a physical structure. *Drosophila* has also been used for population studies in the wild. Flies are set free in known genotypic proportions and their offspring collected later to study changes due to selective forces. The early research linking mutation with ionizing radiation was carried out using *Drosophila*, and the mutagenic effects of many chemical agents were first demonstrated on fruit flies.

## Maize: *Zea mays*

The genetics of corn have been studied most intensely because of its agricultural importance. (Maize also has a claim on the affections of the drinking man since it is the basis of grain whisky which forms the backbone of most blended Scotch whiskies.) The fact that one corn cob represents a large number of offspring, and also the ease of mating the plants as well as the size and shape of corn chromosomes have made it a valuable object of genetic study. The corn plant bears both male and female flowers on the one plant. Like other flowering plants pollination ends in the double fusion of nuclei. One pollen nucleus fuses with two nuclei in the embryo sac of the ovule to form the endosperm nucleus which is triploid. This controls the tissue (the endosperm) which will provide food for the developing embryo. The other pollen nucleus fuses with the egg nucleus to form the zygote which develops into the new plant.

Thus an embryo and its food supply, the endosperm, have the same genes in common, but in a different balance. Using this, one can study dominance relations between alleles in a unique way.

The life cycle of *Zea mays*

## BOOKS TO READ

The books on this list cover most aspects of genetics in rather greater depth than the present work. The last title is included because, although it treats quantitative genetics in a highly technical manner, it is the most recent of the few books devoted to the subject and because it also deals with behavioural genetics.

*Heredity, Evolution and Society* by I. M. Lerner. W. H. Freeman, San Francisco, 1968.

*The Organization of Heredity* by K. R. Lewis and B. John. Edward Arnold Ltd., London, 1970.

*Animal Species and Evolution* by E. Mayr. Oxford University Press, London, 1963.

*Mankind Evolving* by T. Dobzhansky. Yale University Press, New Haven and London, 1962

*Applied Genetics* by D. Paterson. Aldus Books, London, 1969.

*Biometrical Genetics* 2nd Ed. by K. Mather and J. L. Jinks. Chapman and Hall Ltd., London. 1971.

# INDEX

Page numbers in bold type refer to illustrations.

Adaptation 121–123, **122–123**
Adenine 16–17, **20**, **21**, **25**
Allele (Allelomorph) 47–56, 112, 125–127, **126**
Amino acid 20, 22, 24, **26**, **27**, 52, 54, **55**
Anaphase **28**, 29, **30–31**
Animal breeding 136–137, 138–140, **138**
Antibody 86–93, 118
Antigen 86–94
Artificial insemination 136, 138
Autoradiography 19, 40, 149
Autosome 36, 57, 65

Back cross 48, 68
Bacterial conjugation 84, 146–147, **147**
Bacterial transformation **13**, 14, **15**
Bacteriophage **14**, **16**, **17**, **18**, 144–146
Bivalent 33, 34, 36, 39, 40
Blood group 86–93

Cancer 103
Carcinogen 103
Carrier 50, 56, 60, 62–63, 108
Cell 10, **10**
Cell division 10, 11, 28–35
Centriole 29
Centromere 29, 33, **43**
Chiasma **30**, 33, **70**, 71–72
Chloroplast 83
Chromatid 19, **24**, 29, 33, 34, 39, **70**, 71
Chromosome 10, 11, **11**, **12**, 19, 29–43, 57–67, 69–73, 74, 97, **98**, 101, 103, 142, **146**, 147, 149, **153**
Chromosome aberrations 35–43, 78, 97, **98**, 101, 103, 155
Cis-trans complementation **72–73**, 74
Cistron 52, 74
Co-dominance 51
Colour-blindness 59–62, **63**
Compatibility 88, **89**, **90**, 93
Criss-cross inheritance 64
Crossing over *see* Chiasma
Cytogenetics 38
Cytoplasm 10, **10**, 82, 84
Cytoplasmic inheritance 82–85
Cytosine 16–17, **20**, **21**, **25**

Deletion 40, **41**, **42**, 97, 101, 155
Deoxyribose nucleic acid (DNA) 14, **15**, 16–26, 32, 52, 82, 84, 97, **99**, 100 102–103, 144, 145, 147
Diploid 11, 14, 34, **39**, 47, 78

Disease resistance 142
Dominance 49–56, 59, 99, 121, 155
Dosage compensation 67
Double helix **12**, 16, 17
Drosophila 68, **69**, **96**, 103, **149**, **150**, 151 152–155, **153**
Drug resistance 84
Duplication 155
Dwarfism 46–49, 99

Endoplasmic reticulum 12, 21
Enzyme 12, 20, **27**, 54–56, 80, **81**
Enzyme induction 80
Episome 84
*Escherichia coli* 146–149
Evolution 111, 113, 122–123

Fitness 113
Founder effect 114

Gamete 11, 14, **30–31**, 32–34, **35**, 39, 40, 42, 48–49, 68, 72
Gene 47–56
  action 75–85
  expression 75, 76
  frequency 104–123
  penetration 76
  pool **109**, 112, 122–123
Genetic code 22, 24–26, **26**, **27**, 52–55, 74, 97, **99**, 100, 102
Genetic engineering 8, 142
Genetic equilibrium *see* Hardy-Weinberg Law
Genetic predisposition 134, **135**
Genotype 45–50, 125
Genotype frequency 104–105, **104**, **105**, **106**, 108, 111
Glutamic acid 52
Graft *see* Transplantation
Guanine 16–17, **20**, **21**, **25**

Haemoglobin 51–55, **56**
  A 52, 54, **54**, **55**, **56**
  S 52, 54, **54**, **55**, **56**
  E 54, **54**, **55**
Haemophilia 62, **64**
Haploid 11, 14, **31**, 34, 48, 73, 78, 151
Hardy-Weinberg Law **106**, **107**, 108–112
Heritability 131–132
Heterokaryon **9**, 151, 152
Heterozygote 48–50, 59, 68, 90, 108, 113, **116**, **117**, 118, 119, **119**, 120–121, 140
Histocompatibility **92**, 94
Homozygote 48–50, 59, 62, 90, 108, 121
Hybrid vigour 121, 138, 140

157

Immunity 86–95
Immunization 93
Immuno-suppressive drugs 95
Independent assortment 68–69
Interphase **28**, 29, 32
Inversion 40, **41**, **42**, 97, 101

Karyotype 11, **32**, 35–36, 40, 78

Linkage 33, 68–74, 155
   see also Sex-linkage
Locus 47, 57, 68–73, 125–127, **126**, 132
LSD 103
Lyon hypothesis **66**, 67
Lysine 54

Maize 142, **154**, 155
Malaria **119**, 120
Meiosis **30–31**, 32–35, 39, 42, **42**, 48, 57, **70**, 71, 97, 102, 150, 151
Mendel 44–47, 125, 138
Messenger RNA 21–25, **25**, **27**, 80, 82, **83**
Metaphase **28**, 29
Mitosis **28**, 29–32, 39, **39**, **43**, 102
Mongolism 8, **36**, **37**, 38–43, 78
Multifactorial inheritance 124–137
Mutagen **98**, 99, **99**, 102–103, **102**
Mutant 80, 81, 82, **96**, 97, 99, 100, 112, **112**, 113, 122, 138
Mutation 97–103, 112, 116, 121, 123, 155
   see also Mutant
Mutation rate 99–100, **100**, **112**

Natural selection **112**, 113, **114**, **115**, 116–123
Neurospora **148**, 150–151
Non-disjunction 35, 39, **39**, 78
Nucleic acids **12**, 14–27, 101
Nucleus 10, **10**, 11, **12**, 29, **34**

Operator gene 80
Operon 80–82, **81**

Phage see Bacteriophage
Pharmacogenetics 115
Phenotype 45–56, 78, 80, 125
Plant breeding 140–142, **141**
Plasma 86
Pleiotropy 78, **79**
Polygene 135
Polymorphism 101, 120–122, 136
Polypeptide **27**, 52, 74
Polyploid 78, 97
Population genetics 104–123
Porphyria 114–116
Progeny testing 136–137, **137**, 140

Prophase **28**, 29, **30–31**
Protein synthesis **12**, 20–26, 52, 54, 56, 143, **143**
Purine 16
Pyrimidine 16

Radiation 98–99, **100**, 101–103, **102**, 138, 142
Recessiveness 49–56, 59, 63, 108, 113, **114**, **115**, 121
Recombination 34, 35, **70**, 71–72, **71**, 74, 147
Regression 128, **129**–130
Regulator gene 80
Replication 24, **25**, 32
Repressor gene 80
Rhesus factor 90–93, **91**, 118–119
Ribose nucleic acid (RNA) 14, 21–25, 32
Ribosome **12**, 21, 22, **27**

Serum **86**, **87**, 89
Sex-chromatin **34**, 36–37
Sex-chromosome **35**, 36, 37, 40, 57–67
Sex limitation 65–66, **65**
Sex-linkage 57–67, 152
Sickle-cell anaemia 51–54, 56, **118**, 119–1 119–120, **120–121**
Sickle-cell trait 51–54, **56**, **119**, 120
Somatic cell genetics 7
Species formation 123
Spindle 29
Structural gene 80

Telophase **28**, 29
Tetraploid 78
Thymine 16–17, **20**, **21**, **25**
Tissue culture 74
Transcription 24, **25**, 32
Transfer RNA 22, 24, **26**
Translocation 40, **41**, 42, **42**, 101, 142
Transplantation 93–95, **94**, **95**
Triplet 22–25, **26**, 52–54, 97
Tumour 100

Ultra-violet radiation see Radiation
Uracil 21

Variation 35, 44–46, 97, 101, 121
   continuous **44**, 45, **45**, 125–131 132, 133
   discontinuous **44**, 45, **46**, 125
Variegation 83–84
Virus 14, **17**, 144–146, **144**
   see also Bacteriophage

Wheat 140–142, **141**

X-chromosome **35**, 36, 37, 40, 57–59, **60**, 64, **66**, 67, 69, 72
X-linkage 57, 59–63
X-ray *see* Radiation
X-ray diffraction 16

Y-chromosome **34**, **35**, 36, 37, 40, 57–59, 60
Y-linkage 58, 59, **60**, **61**

Zygote 49

## SOME OTHER TITLES IN THIS SERIES

- Arts
- Domestic Animals and Pets
- Domestic Science
- Gardening
- General Information
- History and Mythology
- Natural History
- Popular Science

**Arts**
Antique Furniture/Architecture/Clocks and Watches/Glass for Collectors/Jewellery/Musical Instruments/Porcelain/Victoriana

**Domestic Animals and Pets**
Budgerigars/Cats/Dog Care/Dogs/Horses and Ponies/Pet Birds/Pets for Children/Tropical Freshwater Aquaria/Tropical Marine Aquaria

**Domestic Science**
Flower Arranging

**Gardening**
Chrysanthemums/Garden Flowers/Garden Shrubs/House Plants/ Plants for Small Gardens/Roses

**General Information**
Aircraft/Arms and Armour/Coins and Medals/Flags/Guns/Military Uniforms/National Costumes of the world/Rockets and Missiles/ Sailing/Sailing Ships and Sailing Craft/Sea Fishing/Trains/Veteran and Vintage Cars/Warships

**History and Mythology**
Age of Shakespeare/Archaeology/Discovery of: Africa/The American West/Australia/Japan/North America/South America/Myths and Legends of: Africa/Ancient Egypt/Ancient Greece/Ancient Rome/ India/The South Seas/Witchcraft and Black Magic

**Natural History**
The Animal Kingdom/Animals of Australia and New Zealand/ Animals of Southern Asia/Bird Behaviour/Birds of Prey/Butterflies/ Evolution of Life/Fishes of the world/Fossil Man/A Guide to the Seashore/ Life in the Sea/Mammals of the world/Monkeys and Apes/Natural History Collecting/The Plant Kingdom/Prehistoric Animals/Seabirds/Seashells/Snakes of the world/Trees of the World/Tropical Birds/Wild Cats

**Popular Science**
Astronomy/Atomic Energy/Chemistry/Computers at Work/The Earth/Electricity/Electronics/Exploring the Planets/The Human Body/Mathematics/Microscopes and Microscopic Life/Undersea Exploration/The Weather Guide